Modern technology and
its influence on astronomy

Participants in 'Modern Technology and its Influence in Astronomy', a workshop to
celebrate Hanbury Brown's 70th birthday held at the Royal Greenwich Observatory,
24–26 September 1986

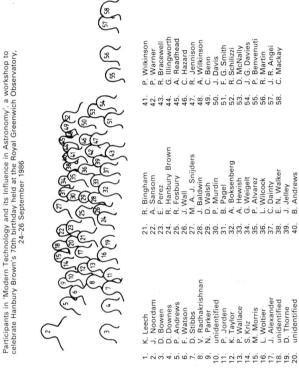

1. K. Leech
2. J. Noordam
3. D. Bowen
4. P. Andrews
5. F. Watson
6. D. Stibbs
7. V. Radhakrishnan
8. N. Parker
9. unidentified
10. P. Jorden
11. K. Taylor
12. P. Wallace
13. S. Kriz
14. M. Morris
15. L. Woltier
16. J. Alexander
17. unidentified
18. D. Thorne
19. unidentified

21. R. Bingham
22. A. Sansom
23. E. Perez
24. R. Hanbury Brown
25. R. Fosbury
26. J. Wall
27. M. A. J. Snijders
28. J. Baldwin
29. D. Walsh
30. P. Murdin
31. B. Pagel
32. A. Boksenberg
33. A. Hewish
34. G. Weigelt
35. P. Alvarez
36. L. Wilcock
37. C. Dainty
38. E. N. Walker
39. J. Jelley
40. B. Andrews

41. P. Wilkinson
42. P. Warner
43. R. Bracewell
44. G. Illingworth
45. A. Readhead
46. C. Hazard
47. R. Jennison
48. A. Wilkinson
49. C. Benn
50. J. Davis
51. F. G. Smith
52. R. Schilizzi
53. D. McNally
54. J. G. Davies
55. P. Benvenuti
56. R. Martin
57. J. R. Angel
58. C. Mackay

Modern technology and its influence on astronomy

EDITED BY

J.V. WALL & A. BOKSENBERG

Royal Greenwich Observatory

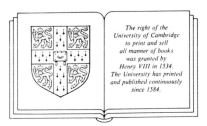

The right of the
University of Cambridge
to print and sell
all manner of books
was granted by
Henry VIII in 1534.
The University has printed
and published continuously
since 1584.

CAMBRIDGE UNIVERSITY PRESS

CAMBRIDGE · NEW YORK · PORT CHESTER · MELBOURNE · SYDNEY

Published by the Press Syndicate of the University of Cambridge
The Pitt Building, Trumpington Street, Cambridge CB2 1RP
40 West 20th Street, New York NY 10011, USA
10 Stamford Road, Oakleigh, Melbourne 3166, Australia

First published 1990

Printed in Great Britain at the University Press, Cambridge

British Library cataloguing in publication data

Modern technology and its influence on astronomy.
1. Astronomical instruments
I. Wall, J.V. (Jasper Vivian), *1942–*
II. Boksenberg, A. (Alexander), *1936–*
522′.2

Library of Congress cataloguing in publication data

Modern technology and its influence on astronomy.
Includes index.
1. Astronomical instruments–Congresses. 2. Brown,
R. Hanbury (Robert Hanbury)–Congresses. I. Wall, J.V.
II. Boksenberg, Alec.
QB84.5.M63 1989   522.2   88-25828

ISBN 0 521 34313 5

TM

Contents

Contributors

J.R.P. Angel, Steward Observatory, University of Arizona, Tucson AZ 85721, USA

J.E. Baldwin, Cavendish Laboratory, Madingley Road, Cambridge CB3 0HE, UK

A. Boksenberg, FRS, Royal Greenwich Observatory, Hailsham, BN27 1RP, UK

J.C. Dainty, Blackett Laboratory, Imperial College, London SW7 2BZ, UK

J.G. Davies (deceased) c/o Nuffield Radio Astronomy Laboratories, Jodrell Bank, Macclesfield, Cheshire SK11 9DL, UK

J. Davis, Chatterton Astronomy Department, School of Physics, University of Sydney, NSW 2006, Australia

D.W. Downes, Institut Radioastronomie Millimetrique, Voie 10 Domaine Universitair, F-38406 St Martin d'Heres, France

F. Graham Smith, FRS, Nuffield Radio Astronomy Laboratories, Jodrell Bank, Macclesfield, Cheshire SK11 9DL, UK

R. Hanbury Brown, FRS, White Cottage, Penton Mewsey, Andover, Hampshire SP11 0RQ, UK

C. Hazard, Institute of Astronomy, Madingley Road, Cambridge CB3 0HA, UK

A. Hewish, FRS, Cavendish Laboratory, Madingley Road, Cambridge CB3 0HE, UK

G. Illingworth, Space Telescope Science Institute, 3700 San Martin Drive, Baltimore MD 21218, USA

P.R. Jorden, Royal Greenwich Observatory, Hailsham, BN27 1RP, UK

A.C.B. Lovell, FRS, Nuffield Radio Astronomy Laboratories, Jodrell Bank, Macclesfield, Cheshire SK11 9DL, UK

C.D. Mackay, Institute of Astronomy, Madingley Road, Cambridge, CB3 0HE, UK

J.E. Noordam, Radiosterrenwacht, Postbus 2, 7990 AA Dwingeloo, The Netherlands

V. Radhakrishnan, Raman Research Institute, Bangalore 560 080, India

A.C.S. Readhead, Owens Valley Radio Observatory, California Institute of Technology, Pasadena 91109, USA

R.T. Schilizzi, Radiosterrenwacht, Postbus 2, 7990 AA Dwingeloo, The Netherlands

K. Taylor, Anglo:Australian Observatory, Epping, NSW 2121, Australia

P.T. Wallace, Rutherford Appleton Laboratory, Chilton, Didcot, OX11 0QX, UK

G. Weigelt, Physikalisches Institut, Erwin-Rommel-Strasse 1, D-8520 Eriangen, West Germany

L. Woltjer, European Southern Observatory, Karl-Schwarzschild-Str 2, D-8046 Garching bei Munchen, West Germany

S.P. Worswick, Royal Greenwich Observatory, Hailsham, BN27 1RP, UK

C.G. Wynne, FRS, Royal Greenwich Observatory, Hailsham, BN27 1RP, UK

Preface

Robert Hanbury Brown was 70 on 22 August 1986. To celebrate this anniversary of a distinguished colleague, we held a symposium in his honour on 23–25 September 1986 at the Royal Greenwich Observatory. For want of a better title we termed this 'Modern Instrumentation and its Influence on Astronomy'. The contributions of necessity were somewhat selective; today a full-scale symposium would have difficulty doing justice to such a topic. There was a blend of surveys, background, and foreground to some of the key and exiciting developments in instrumentation with emphasis on areas of astronomy to which Hanbury has made such major contributions. At Hanbury's request, it was forward-looking rather than backward-looking. It highlighted the remarkable synthesis – if we may use the word – of techniques in radio and optical astronomy towards high resolution. It provided a forum for the gathering of Hanbury's friends, distinguished scientists pre-eminent in their fields, whose reminiscences of the past and animated discussions of the present and future ran past the appointed sessions and well into the night. It was a tribute to a remarkable man that so many distinguished astronomers and physicists came from so far for the occasion. The proceedings were a treat for the participants, an education for all who attended, and an eye-opener in particular for the current crop of students at the University of Sussex-with the RGO.

We hope that collecting the contributions which all the distinguished participants were good enough to prepare will convey some of the flavour of this meeting. It is a further tribute to the guest of honour that all the speakers did so in the face of their heavy commitments. At his request, any proceeds which accrue from the sale of this volume will support young astronomers in the UK.

It was fun, Hanbury. Many happy returns. Do you and Heather have plans for your 80th birthday? Can we book you for 1996?

J.V. Wall
A. Boksenberg

xi

1

Robert Hanbury Brown

A.C.B. LOVELL

We are assembled to pay tribute to one of our distinguished colleagues and my pleasant task is to open the proceedings by saying a few words about Hanbury Brown. Since I have known him for nearly half a century that ought to be an easy thing for me to do. In fact, the prospect alarms me, mainly because only once before have I had to pay tribute to a 70-year-old in his presence and afterwards his wife thanked me for preparing such an excellent obituary. Fortunately the detailed accounts of Hanbury's impact and influence on astronomy will be the task of other speakers and I simply want to give you some background.

In view of the subject of this symposium it is understandable that most of you here today think of Hanbury Brown in connection with his stellar interferometer in Australia and as a Professor of Astronomy recently retired from the University of Sydney. Maybe there are a few who might remember that he once worked at Jodrell Bank. Whether he is to be regarded as an Australian now visiting England, or as an expatriate Englishman whom we are now glad to welcome back to this country, or as a national of India, who has spent his research career in England, America and Australia, I must leave him to elucidate.

Whatever these uncertainties may be there is no doubt that he began his research career in England – and not at Jodrell Bank either. His first important research was concerned with the polarization of radio waves but this work had no connection whatsoever with astronomy. I want to read a short paragraph from a paper Hanbury Brown dated 1938 which is to be found in the Churchill College archives.

It was necessary to decide which polarization should be used. The effect of polarization was investigated by fitting an aircraft with both horizontal and vertical

1

aerials, and also by violently altering the flight attitude of an aircraft carrying a fixed aerial system. It was found that the detection ranges of ships were similar with both horizontal and vertical polarization, but the scattered energy which was reflected by the sea was less with horizontal than with vertical. For this reason horizontal polarization was adopted for ASV.

ASV was the radar code for air to surface vessel and in its successive forms played a vital part in the battle of the Atlantic. That pre-war experiment of Hanbury Brown's was probably the most important he has ever carried out. It determined the nature of the aerials on all of the ASV equipments that operated with Coastal Command until the advent of the centimetric ASV in 1942. Even with the long range versions, that is those fitted with broadside arrays on Wellington aircraft or Sunderland flying boats, the range of detection on a surfaced submarine was only about 8 miles. If Hanbury Brown had made the wrong decision about polarization that range in a rough sea would have become negligible and the consequences for our merchant fleet would have been incalculable.

You might think that this experiment was an easy one to carry out. In the roomy jet propelled aircraft of today it probably would be an enjoyable investigation to make. Things were different in the 1930s. Probably Hanbury carried out that research either in a Heyford bomber or an Anson and it was a hazardous enterprise to make the violent changes to the flight attitude which he described in that paper. But that was the direct and no-nonsense approach to observational problems that enabled him to make such an instant impact on radio astronomy a decade later.

The chances that I became associated again with Hanbury after the War must have been very slender indeed. It is not my intention to trace his career during the War and immediately afterwards. In fact, he had gone to America and we were out of touch and he alone can tell you the motives that led him to write to F.C. Williams who was then the Professor of Electrical Engineering in the University of Manchester. One afternoon early in May 1949 Freddie Williams phoned me at Jodrell where a few war-time trailers served as our laboratories. Williams was then engaged on his brilliant early digital computer developments but his telephone call had nothing to do with this work. The message which he conveyed to me was that he had received a letter from Hanbury Brown who wished to explore the possibility of returning temporarily to a university to do research for a PhD. He generously suggested that my developments at Jodrell might be of more interest to him than his own work. The suggestion seemed like a gift from heaven. A few weeks later Hanbury came to see me. With Blackett's help the problem of a Fellowship was soon resolved and in September 1949 Hanbury joined the small group at Jodrell.

It would be difficult for me to exaggerate the significance of his arrival. His wide experience of the world, and his scientific and technical authority made an instant impression on Blackett and all members of the scientific establishment with whom we had contact. Blackett, the senior members of the University administration, and the external sources from whom we were demanding ever-increasing sums of money no longer had to base their judgement on my opinion only of what should be done. If they queried my extravagant demands they could ask Hanbury Brown and we soon began to make the kind of progress that would otherwise have been impossible.

This progress concerned a device that we called the transit telescope. Towards the end of 1947 we had completed the construction of a paraboloidal reflector 218 ft in diameter. We made it out of scaffolding tubes, wire hawsers, and 16 miles of thin wire. The focus was at the top of a 126-ft tubular steel mast. We had built this in order to get the sensitivity which I thought would be necessary to obtain radar echoes from high energy cosmic ray air showers – which is the reason why I had taken the trailers of war-time radar equipment to Jodrell. However, as it transpired eventually, in the calculations made during war-time duties, I had made an error of about a million times and we did not find the types of echo for which we were searching. Since the instrument had a very narrow beam by the standards of those days I had suggested to Victor Hughes, who had joined me as a research student, that he might use the telescope to study cosmic noise (as the phenomenon was then called). By the time of Hanbury's arrival in September 1949, Hughes had produced his MSc thesis on some characteristics of the radio emission from this zenithal strip of the sky. More than 30 years later Hanbury himself described the sequence of events that followed. He wrote this:

...They had built, with their own hands, a fixed paraboloid with the astonishing size of 218 ft. The original purpose of this remarkable dish had also been to detect cosmic rays by radar, but by the time of my first visit to Jodrell in 1949 they had given up that programme and realized that, by happy accident, they had built a powerful instrument with which it should be possible to study the radio emissions from space. Lovell suggested that I should join them and take charge of a 'cosmic noise' programme which had already been started by Victor Hughes. With Reber's paper on 'Cosmic Static' in mind I accepted his offer with enthusiasm and went to work at Jodrell Bank in September 1949. Shortly afterwards I was joined by a research student, Cyril Hazard, with whom I was to work for several years... As a first step we chose to work at the highest frequency at which the antenna could be expected to perform reasonably well; we wanted the narrowest possible beam. To that end we removed the original primary feed of the paraboloid, which had been designed to work on 4.2 m with a high power transmitter using open-wire feeders; in its place we substituted a primary feed designed for 1.89 m and connected it to the laboratory by 300 ft of coaxial cable with a loss of about 2.7 db. I well remember soldering this coaxial cable to the primary feed while perched on the top of the 126 ft tower at the centre of the paraboloid; it was not a job which I would care to do again...

After describing how they built a sensitive and stable receiver for this wavelength, Hanbury continued

...we were able to make most impressive records of the 'cosmic noise' power received from the zenith within a 2° beam. I shall never forget the thrill of seeing those first records. They showed clearly the broad maximum as the galactic plane passed through the aerial beam; they also showed a few discrete sources or, as we called them in those days, radio stars.

Then, in the early months of 1950, Hanbury Brown and Hazard took a vital step – they decided they could direct the beam away from the zenith by tilting the mast.

...The beam could, of course, only be swung by displacing the primary feed and, in practice, that meant by tilting the central mast. The mast itself was far from self-supporting; it was made by slotting 6 sections of tubular steel together and supporting them by 18 guy wires. The bottom section was hinged at the ground so that it could be tilted in a north-south plane. We set up two theodolite stations east and west of the mast and, very gingerly, tried tilting the mast by making successive adjustments to the 18 guys. It was a very anxious business; the tilting had to be done in almost imperceptible stages so as not to kink the mast and it took Hazard and myself about two hours running from guy to guy and shouting at each other, to tilt the beam through one beamwidth (2°). It was quite good fun on a fine day, although it was a rather slow way of scanning the sky. The early days of any new science are apt to be laborious. The largest tilt at which we could persuade ourselves that the mast was safe and that the beam shape was still respectable was 15°. This meant that we could survey a strip of sky between declinations + 38° and + 68°... In those days very little was known about cosmic radio noise and our paraboloid was larger and had a much narrower beam than any other single antenna that had been used to scan the sky. Our programme, therefore, was simple and exciting: it was to look at everything in our field of view...

In 1950, when Hanbury Brown and Hazard commenced using the transit telescope, Ryle and his colleagues in Cambridge had discovered 50 discrete radio sources and the Australians had published a catalogue of 22 similar radio sources. The remarkable fact was that, apart from the identification of one of the radio sources with the Crab Nebula, not one of the discrete, localized, radio sources could be positively identified with any of the stars or galaxies in the catalogue of the optical astronomers. The distribution of these radio sources appeared to be isotropic and it was widely believed that they must be a hitherto undiscovered type of dark 'star' in the solar vicinity. They became generally known as radio stars and at that stage in 1950 the firm belief was that the phenomenon of the cosmic radio waves was localized to the Milky Way system. Fortunately M31 was within the strip of sky that could be mapped by tilting the mast to the extreme limits. An elementary calculation suggested that if the Milky Way were to be placed at the distance of the Andromeda galaxy, then it should be detectable by using the transit telescope. At that stage in 1950, when almost without exception, astronomers believed that the cosmic radio waves so far studied were generated in our own galaxy, this appeared as a most vital question.

   These observations could only be made when the Andromeda nebula was in transit in the middle of the night since complete freedom from other sources of noise – whether solar or man made – was absolutely essential if the necessary sensitivity was to be realized. Hanbury Brown and Hazard began this survey of M31 in August 1950. When they were satisfied with the record from one narrow two deg strip of sky the arduous process of changing the tilt of the tower had to be carried out. They were nearly driven to despair by a sequence of thunderstorms in the middle of the night. It took 90 nights to make the contour maps and the survey was completed just before the vital region of the sky moved out of the radio-quiet night period when in transit. Their results established beyond doubt that the Andromeda nebula also emitted radio waves of much the same intensity as the local Milky Way galaxy. By the standards of the day the work was a *tour de force*. Radio waves had been detected from a nebula 2 million light years distant and it was no longer possible to regard the local galaxy as in any way unique as a radio emitter. However, in a curious way the observations deepened the mystery about the identity of the localized sources already catalogued by the groups in Sydney and Cambridge. Many of these sources were strong radio emitters – they had been discovered by using small aerial systems of low gain. But even with the great sensitivity of the transit telescope the radiation from Andromeda was difficult to detect – it was a weak source of radio emission compared with the majority of discrete sources already discovered. For this reason Hanbury Brown and Hazard reached the conclusion that these unidentified sources could not be extragalactic. They were too strong and must, therefore, be 'radio stars' in the local galaxy – an interesting reminder that in those days no one had seriously considered the possible existence and influence of magnetic fields or of non-thermal processes in the Universe.

It is not my intention to pursue the historical sequence further or give the details of Hanbury Brown's involvement with the development and use of the Jodrell long baseline interferometer. Graham Smith and J.G. Davies will speak about these topics, but there are two other matters I must mention.

The first concerns the impact of this work of Hanbury Brown on the idea of the large steerable radio telescope that was eventually built as the 250-ft Mark I. Many months before Hanbury arrived at Jodrell I had been talking to Blackett about my desire to build a steerable version of the transit telescope and had, in fact, already made discouraging contacts with various engineering firms about such a possibility. Of course, that work of Hanbury Brown and Hazard on M31 was the perfect practical demonstration needed to support my arguments about the importance of a steerable narrow beam radio telescope. In the face of those results the desirability and value of such an instrument could no longer be questioned and the issues became the practical ones of finance and engineering – but these are outside the scope of this talk.

The second matter concerns the development that led to Hanbury Brown's departure from Jodrell in 1962. I have frequently read references to the intensity interferometer implying that it was an idea of Hanbury's which he initiated in Australia. In fact, this brilliant development occurred whilst Hanbury was at Jodrell and I am glad of this opportunity to summarize the sequence of events. I have already referred to the problem of the discrete sources, or radio stars, of which about 75 had been catalogued in Cambridge and Sydney at the time when Hanbury Brown and Hazard made their measurements on M31 in 1950. Graham Smith had made the positional measurements which enabled Baade and Minkowski to identify the sources in Cassiopeia and Cygnus but neither the Sydney nor Cambridge interferometers had been able to resolve the radio sources, and the sole information was that the angular diameter of the radio source in Cygnus was less than about 8 arcmin. Of course, this question was of fundamental importance – we did not know whether the radio sources had angular diameters typical of the stars or were tens of thousands of times larger – that being the limit of the existing radio measurements.

The maintenance of phase stability with interferometer aerials separated by more than a few tens of kilometres had not then been solved, neither was there sufficient sensitivity to deal with the problem by reducing the wavelength. It was this dilemma that stimulated Hanbury Brown to his idea of the intensity interferometer in which it is not necessary to maintain the phase of the signals in the separate aerials up to the point of combination, and hence no limit need be set to the length of the baseline. At that time the transference of this idea into a practical system presented a severe technical problem. Fortunately Roger Jennison, who had been a war-time aircrew navigator, came to Jodrell as a research student in the autumn of 1950. With his electronic knowledge and the help of a young visiting Indian student M.K. das Gupta, Hanbury was able to transfer the idea to a practical system to the extent that the correctness of the theoretical idea was demonstrated by measuring the solar diameter in the summer of 1951. A year later, Jennison and das Gupta had refined this prototype and measured the angular diameters of Cassiopeia and Cygnus. These were the first measured angular diameters of radio sources to be published but, as you will be aware, simultaneously a relatively small extension of the baseline of the conventional radio interferometer had given similar results both in Cambridge and Australia. Hanbury had gone to Australia for the URSI meetings and had learnt about these latter results of B.Y. Mills whilst there. He wrote to me: 'Apparently most sources are quite large. It seems that we may have designed our interferometer to crack a nut which could have been more easily opened.'

Maybe, but although the intensity interferometer had little consequence in the mainstream development of radio astronomy it was to have a memorable impact in unforeseen directions. Whilst the radio intensity interferometer was being built, Richard Twiss was a frequent visitor to Jodrell and it seemed to me that he was far more interested in the development of the theoretical ideas of the device than in his work at the Services Electronics Research Laboratory at Baldock. In 1954 he and Hanbury published an account of the interferometer and the last paragraph referred to a subject which was by that time monopolising most of the discussions at Jodrell. I will quote that paragraph:

The use of the 'Michelson' interferometer at radio wavelengths is a logical extension of optical practice and it is interesting to enquire whether the principle of the new type of interferometer can in turn be applied to visual astronomy, since in this way it might be possible to increase the resolving power and mitigate the effects of atmospheric turbulence. A preliminary examination of this question, which will be discussed in a later communication, suggests that the technique cannot be applied to optical wavelengths, and that it breaks down due to the limitations imposed by 'photon noise'.

The possibility of applying this new technique in optical astronomy opened dramatic possibilities. For a long time the arguments between Hanbury Brown and Twiss continued as to whether the transference of the system to the visual domain would be possible. Caution was engendered by an apparent conflict with fundamental physical theory. In order to use light waves the two radio telescopes would have to be replaced by two mirrors with photoelectric cells as detectors. The output of the photo-cells would then have to be brought to a correlator as in the radio case. Then, as a function of the separation of the mirrors, the correlation between the fluctuations in the currents from the cells would be measured when the mirrors were directed at a star.

There were at least three major and perhaps decisive objections to this concept. First, it was uncertain whether atmospheric problems might inhibit the correlation. Second, it was essential to the operation of the system that the time of arrival of photons at the two photo-cathodes should be correlated when the light beams, incident on the two mirrors, were coherent. This effect had never been observed with light, and indeed some theorists we consulted maintained that the effect could not exist because it would be in violation of fundamental physical theory. Thirdly, even if the effect did exist it was not clear that the correlation would be fully preserved in the process of photo-electric emission.

These last two fundamental issues were the subject of a laboratory experiment at Jodrell Bank during 1955. The apparatus was constructed in a small room which had been built and used for a short time to house a spectrohelioscope. The light source was a mercury arc and the two coherent light beams were obtained by division at a half-silvered mirror. The separate light beams were fed from the half-silvered mirror to the two photo-multiplier tubes along paths at right angles. The fluctuations in the output currents from the two cells were multiplied together in a correlator. The average value of the product recorded on the revolution counter of an integrating motor gave a measure of the correlation in the fluctuations.

The results were first published early in January of 1956. At the end of their account, the authors wrote: 'This experiment shows beyond question that the photons in two coherent beams of light are correlated, and that this correlation is preserved in the process of photoelectric emission. Furthermore, the quantitative results are in fair agreement with those predicted by classical electromagnetic wave theory and the correspondence principle.'

The publication of this paper caused a storm. People elsewhere carried out experiments to show that the interpretation of the results was erroneous and several eminent theorists maintained that the conclusions were in violation of fundamental quantum theory.

It was against this background that Hanbury Brown and Twiss built a prototype version of this optical intensity interferometer to make a practical test of the system on Sirius. Two ex-army searchlights on their mounts were borrowed to serve as the mirrors. They were placed several yards apart outside the window of the new control room for the radio telescope. At that stage the room was almost empty of the control equipment for the telescope, and the recording equipment attached to the searchlights was placed in this room. The observations were made in two stages. In November and December 1955, and then with increased spacings during January to March 1956. The investigation was completely successful and the diameter of Sirius was measured as 0.0063 arcsec. The first practical measurement of a stellar diameter for nearly 30 years had been achieved.

This brilliant and unexpected diversion of the radio interferometer technique was soon to have consequences which none of us foresaw. Naturally, having made the prototype observations Hanbury Brown proceeded to design and build a much larger device with mirrors 23 ft in diameter, moving on a circular railway track of 600 ft diameter. The need for clearer skies than those of Cheshire, the desire to measure stars of the Wolf-Rayet type visible in the Southern hemisphere, and the circumstance that Twiss had moved to Australia, led eventually to the shipment of the new interferometer to New South Wales, where it was erected in a suitably remote spot at Narrabri. Hanbury Brown who had been appointed to a personal chair of radio astronomy in 1960 was given leave of absence in 1962 by the University of Manchester. Alas, the leave extended from one year to two, and then to a final break when he decided that in fairness to all concerned he ought to accept the offer of a chair in Sydney and remain there to use his new interferometer – which he did with brilliant success.

Those of you who visited Narrabri will be well aware of the immense practical difficulties that Hanbury encountered in setting up this interferometer in that remote place. Shortly after his arrival in April 1962 he wrote to me about the problems of working at Narrabri: 'Most people laugh heartily when one mentions distance as a difficulty as they regard 400 miles as a reasonable distance from a shopping centre, but I have not yet reached that desirable view point.' Early the next year I received a handwritten letter from Cyril Hazard dated 9 February 1963. He had gone to Narrabri to help Hanbury, and this letter complained about the temperature of 104 °F in which he was expected to work and about the problems of living and money.

The fact that during a visit to Parkes, Hazard had made the classic measurements which led to the identification of the quasar 3C 273 merely appeared as a postscript to the agonies of Narrabri.

After many delays Hanbury eventually received the correlator from England and was able to make some tests with the interferometer. On 25 June 1963 he wrote to me about some technical problems. 'The correlator has exhibited some really nasty faults. It has digital data handling and I would support a programme for murdering everyone concerned with digital data handling so that the knowledge of this technique would become extinct...If we could afford it I would take all the sophisticated data handling equipment and roll it out flat with a steam roller.' It is a great example of Hanbury's scientific single-mindedness that he surmounted high temperatures, floods that blocked roads, insects of all descriptions, lizards on the railway track and ants that ate the laboratory to produce a series of astronomical results of great elegance and importance.

However, Narrabri came to an end when every possible result had been squeezed out of that equipment and I believe that John Davis will describe to us tomorrow morning what then happened. I saw the successor to Narrabri in the grounds of the Australian National Measurement Laboratory in Sydney in 1985 and thought that only Hanbury would be capable of using the resources of modern technology to create such a brilliant reversion to classical interferometry.

Hanbury's biographer will have to write about many other things in his life. Of his work on radar of which I have given only one example, of his excursion into the fields of industry and commerce after the war, of his Presidency of the IAU and of his books. For myself I am glad that he had the urge to study for a PhD in 1949 – although I might add that he is one of the few Jodrell candidates who never got the degree. Greater things intervened.

2

Interferometers and aperture synthesis

F. GRAHAM SMITH

Interferometers were originally used in radio astronomy to distinguish discrete radio sources from the diffuse background. A small diameter source, such as the Sun, responded to the interference pattern of an interferometer pair, while the galactic background did not. Even if Michelson had not already shown that an optical interferometer with variable spacing could be used to measure the diameter of a star, it was an obvious idea to use radio interferometers with variable spacings to measure the diameters of the newly discovered Cygnus A and Cassiopeia A.

What spacings would be required? A reasonable guess suggested that for metre wavelengths some kilometres would be required, and the radio-linked interferometer was therefore developed both at Jodrell Bank and in Australia. Accounts of these pioneering long-baseline interferometers have been given by Hanbury Brown and by Mills in Sullivan's *The Early Years of Radio Astronomy* (1984). It happened that the discrete radio sources known at that time had unexpectedly large diameters, so that their diameters were in the event measured by cable-linked interferometers. The radio-linked interferometer had been shown to be practical, both in conventional form and as an intensity interferometer, and the technique was developed by Hanbury Brown and by Palmer at Jodrell Bank, leading to the isolation of a class of very small-diameter sources, the quasars.

So far the story is one of analysis rather than synthesis. Michelson recognized that the synthesis of the profile of a spectral line, or of the brightness distribution across a star, could not be achieved unambiguously from fringe visibilities alone. The simple Fourier relationship between the required brightness distribution and the correlation structure of a wavefront was becoming clear even by 1954 (Bracewell & Roberts, 1954), but there were severe technical problems in the practical realization of synthesis, as noted by Scheuer (Sullivan, 1984). The severest problems were the preservation of phase over large baselines, and the formidable amount of computation required for the transformations. I recall in the early days using the Lipson–Beevers strips developed by X-ray crystallographers for a painstaking point-by-point numerical synthesis. The first synthesis telescope, which was the transit Tee built in Cambridge by Blythe, was only possible because of the availability of EDSAC.

The idea of synthesis, as conceived by Martin Ryle and his colleagues, was to observe separately all the interferometer pairs which existed within a large telescope aperture. The explanation of the technique was in terms of a chessboard, on which only two pieces could be placed: successive placings, i.e. successive observations, used all the possible vector spacings between the two pieces. The idea was to recreate a uniformly filled aperture. The Tee synthesis telescope had one such small element, which was moved along a line perpendicular to a long narrow array. The intention was always to sample the whole range of spacings; as we would now say, the whole of the u–v plane.

Rotation synthesis

Martin Ryle's biggest step came when he used the rotation of the Earth to sweep a range of vector spacings with a fixed pair of antennas. The first result was a survey of the North Polar region (Ryle & Neville, 1962). The map showed radio sources down to 250 microJansky (mJy), with a resolution of 4.5 arcmin. It was exhibited at a Herstmonceux conference, and so deeply impressed Professor Jan Oort that the plans for a Benelux Cross soon became plans for the Westerbork synthesis Radio Telescope.

Up to and including the One-Mile Telescope, the 5 km telescope, and the WSRT, there was a clear intention to leave no gaps unfilled in the u–v plane. At the same time, longer and longer baselines were being used at Jodrell Bank for interferometers designed to measure diameters. But these interferometers had several different baselines, and they observed for long periods as the Earth rotated. Could they be used for mapping? Large gaps in the u–v plane and no phase at first made synthesis seem impossible. But that is what MERLIN now achieves.

Undersampled apertures

The first step was taken by Jan Hogbom, who introduced the CLEAN process. An ideal point source observed by an unfilled aperture will be mapped as a point surrounded by large and widespread sidelobes. If it is the only point source the pattern will match the theoretical pattern of the unfilled aperture; therefore subtract the ideal pattern and look at what is left. If another weaker source is present, repeat the process. The result will be a list of point sources and a residual map of random noise. Hogbom pointed out that this would work if, as is usual, most of the map is empty. It worked better than anyone expected, even if most of the source was extended.

The VLA itself began to perform as though it sampled the u–v plane fully, and the VLBI networks started to produce real maps even though the holes in the u–v plane made it look more like a nylon fishing net than a conventional Mk I radio telescope surface.

The remaining problem was calibration in amplitude and in phase of a widely separated and often diverse set of telescopes. Here again the image itself gave the clue to the errors: if the point source response was incorrect, then it might be put right by adjusting the amplitudes and phases of the recorded Fourier components. This process, of adaptive calibration, is now commonly used in aperture synthesis. It was introduced in its present form by Readhead & Wilkinson (1978) and by Cornwell & Wilkinson (1981). A full account is best found from the recent seminal book by Thompson, Moran & Swenson (1986); a shorter historical account is given by Smith (1986).

MERLIN and the EVN

These various threads of development have now been drawn together in the various high-resolution mapping telescopes, which form a sequence, roughly as follows: 5 km, WSRT, VLA, MERLIN, VLBI in Europe and in USA, World-Wide VLBI.

The sequence is evidently in the maximum available spacing, which is also roughly the sequence of angular resolution. It is also a sequence in decreasing filling factor. Technically, there is a break between MERLIN and VLBI, which is the break between connected elements and tape recording. I shall now concentrate on MERLIN, which is physically the largest network of connected elements. Some of these elements are also used for the European VLBI Network (EVN), so that the extension of MERLIN which is now under way has also taken the EVN into account.

MERLIN has at present five telescopes outside Jodrell Bank; only three of these are suitable for short wavelengths. The maximum baseline is 134 km. Improvements already in hand include an extension at Cambridge, giving baselines up to 200 km, including a useful extra east – west component; the intention is to use initially one of the small telescopes of the One-Mile Telescope, and also the 151 MHz array.

Another important improvement concerns the Mk II telescope at Jodrell Bank. A new reflector surface has now been added, using a new simple design of panels. Adjustments have been made after holographic measurements, using Mk IA as the reference antenna. The surface accuracy is better than 500 μm rms over a wide range of elevation angle. We already know that MERLIN will work at 1.3 cm wavelength, i.e. the water maser line, and we now have four antennas for this wavelength. We are confident that a similar improvement can be made to the Mk IA telescope, where we will aim for 1 mm rms. This will involve a jacking system to compensate for gravitational deformation. The rms is about 6 mm at present, with larger deviation in some areas and at large zenith angles.

More immediately we expect to build a new 32 m telescope at Cambridge, with rms accuracy below 1 mm, possibly around 0.5 mm. A design study is already funded.

When these improvements are complete, a remarkable scientific programme becomes available. Some examples are:

(i) In star-forming regions and circumstellar shells, the observations of OH and H_2O masers with an angular resolution seven times that available from the VLA. In addition to the astrophysics of star formation and stellar winds, there will be a direct astrometric interest in the expanding shells which will give new and independent measurements of the distance of the galactic centre.

(ii) Many stars, including especially T Tauri, Wolf-Rayet, O and B stars, are detectable in thermal emission, and others, including especially novae, are detectable in non-thermal emission. In novae the expansion parallax of the emitting cloud should be measurable with MERLIN, while in radio-emitting binary systems the orbits will be resolvable. For example, at 22 GHz the position of a 10^4 K thermal source will be measured to 1 milliarcsec accuracy; combination with optical doppler measurements will then yield the complete binary orbit.

(iii) X-ray binaries, and exotic binaries such as SS433 and Cyg X-3 can be resolved, giving a detailed picture of the collimation of the jets, and the velocity of emitted plasmons.

(iv) In extragalactic studies MERLIN will be combined with appropriate measurements from the VLA and the EVN to produce very much better maps on a wide range of angular scales, including polarization. This will allow mapping of high-z objects, where significant differences from more local objects are expected; mapping of small separation gravitational images; details of beams and jets, and of hot spots in powerful double sources. The active nuclei of Seyfert and other galaxies now appear to be closely related to quasars; these can be studied in the radio continuum and in line emission from OH and H_2O masers. MERLIN is already contributing to studies of Starburst galaxies and megamaser galaxies; the improvements will allow mapping of all known megamasers, with angular resolution down to 10 milliarcseconds. All galaxies with z up to 0.05 become prime candidates for hydrogen and OH line absorption measurements, giving maps of the radial flow of extranuclear material.

(v) Astrometry with MERLIN depends primarily on the longest baselines, which will be most improved by the new Cambridge telescope. Fundamental astrometry now enters a new era, when a grid of galactic star positions can be related to quasar positions, while the pulsars allow a new and direct comparison between ecliptic and celestial coordinate systems. Differential astrometric measurements have already yielded parallaxes for many pulsars; the improved MERLIN will extend these measurements to longer baselines and shorter wavelengths, with an improvement of a factor of four in precision.

A backward glance

It is astonishing when one looks back over a very few decades and sees the transformation in astronomy and especially in radio astronomy. In

1952 I happened to be in the right place at the right time to measure the position of four radio sources to an accuracy of 1 arcmin, and to measure diameters of several arcmin. Soon afterwards the Jodrell Bank work showed that Cygnus A was a double radio source. No-one was clear on the nature of the other discrete sources, or on the background radiation from the Galaxy: indeed these two problems were merged into one, so that the sources were thought of as stars which together must make up the galactic background.

MERLIN is now making maps to a resolution typically of 0.1 arcsec, while VLBI typically reaches 1 milliarcsec. This is 5 decades better than the first diameter measurements. Parallax and proper motion now require positions to a milliarcsec or better: geophysical applications of VLBI similarly work to about 1 cm in 1000 km. I remember in these first positional measurements wondering how to measure an interferometer baseline of only 100 m or so reliably to about 1 cm.

Again, it is astonishing to think that the Jansky was decided on as a convenient unit for the lowest flux density of any real interest: the improved MERLIN, at 5 GHz and at 22 GHz, will reach 100 μJy.

The really significant change is, however, not so much in the engineering but in the astronomy. A radio astronomer *per se* is becoming a rarity, and our expertise is part of the broad range of techniques of modern astronomy.

References

Bracewell, R.N. & Roberts, J.A. (1954). *Aust. J. Phys.*, **7**, 615.
Cornwell, T.J. & Wilkinson, P.N. (1981). *Mon. Not. R. Astr. Soc.*, **196**, 1067.
Hogbom, J. (1974). *Astron. and Astrophys. Suppl.*, **15**, 417.
Readhead, A.C.S. & Wilkinson, P.N. (1978). *Ap. J.*, **223**, 25.
Ryle, M. & Neville, A.C. (1962). *Mon. Not. R. Astr. Soc.*, **125**, 39.
Smith, F. Graham (1986). *Q. Jl. R.A.S.*, **27**, 543.
Sullivan (1984). *The Early Years of Radio Astronomy*, ed. W.T. Sullivan III, Cambridge University Press.
Thompson, A.R., Moran, J.M. & Swenson Jr, G.W. (1986). *Interferometry and Synthesis in Radio Astronomy*. John Wiley & Son.

3

Aperture synthesis and the microwave background

A. HEWISH

Summary

The suggested title for this talk was 'Aperture Synthesis: Past, and Future'. The past has already been covered by the Astronomer Royal in his broad survey; so I therefore propose to concentrate upon the future.

Radio telescopes using aperture synthesis now cover the globe, extending to longer baselines and shorter wavelengths in the quest for higher angular resolution. We have, for example, the VLA, MERLIN, VLBA, global VLBI, QUASAT[†], while millimetre-wavelength arrays exist or are under construction in several countries. At the longest wavelengths a giant instrument is also planned in India. We therefore seem to be well-served in observational phase space by existing or planned systems. So what remains for aperture synthesis in the future? Are there, in fact, any new frontiers to be developed? One challenging field which remains to be exploited is that of small-scale structure in the microwave background radiation. It is of fundamental importance for observational cosmology and the Mullard Radio Astronomy Observatory (MRAO) is planning to work in this area. In outlining some of the possibilities I am reflecting especially the thoughts and aspirations of the younger radio astronomers at Cambridge.

First then, what are the sources to be mapped and what do they tell us about the Universe? Modulation of the background temperature can arise

---

[†] VLA: Very Large Array of the National Radio Astronomy Observatory (MRAO), USA; MERLIN: Multi-Element Radio Linked Interferometer of Nuffield Radio Astronomy Laboratories, Jodrell Bank, UK; VLBA: Very Long Baseline Array, under construction by MRAO, USA; VLBI: Very Long Baseline Interferometry; QUASAT: proposed dedicated VLBI satellite of the European Space Agency.

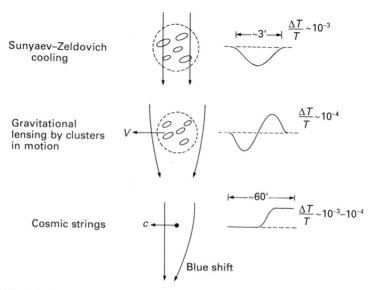

Fig. 3.1. Possible sources of small-scale structures in the microwave background radiation.

from a variety of known or postulated causes, as shown schematically in Fig. 3.1.

(a) Sunyaev-Zeldovich Cooling

Microwave background photons are shifted to higher energy by Compton scattering from electrons in the hot gas within clusters of galaxies. This produces a decrement of temperature ΔT such that $\Delta T/T \sim 10^{-3} - 10^{-4}$ on an angular scale, typically, of a few arcmin. This long-sought effect has now been confirmed in three clusters (Birkinshaw, Gull & Hardebeck, 1984). The decrement is proportional to $N_e T_e$ for the electrons so mapping would give the distribution of gas pressure across the cluster. This information is complementary to X-ray data which measures a different function of density and temperature.

(b) Gravitational lensing by a moving cluster

Gravitational deflection of the background radiation caused by transverse motion of a cluster-mass cloud, for not unreasonable parameters, produces a modulation $\Delta T/T \sim 10^{-4}$ having a different characteristic signature (see Fig. 3.1) on the angular scale of clusters (Birkinshaw & Gull, 1983).

(c) Cosmic strings

If the Grand Unified Theory is correct we might expect the early universe to be laced with strings having masses of ~ 1 galaxy kpc^{-1}, and whiplash speeds $\sim$ c. When strings join to form closed loops, several of which might lie within the horizon at the era of recombination, the motion of their associated gravitational potential wells through the cosmic fluid would generate density perturbations, and hence temperature perturbations, where $\Delta T/T \sim 10^{-3} - 10^{-4}$ in either positive or negative sense. The angular scale is a few arcmin (Chase, 1986).

For strings which extend from horizon to horizon a different effect can occur (see Fig. 3.1). Space-time curvature causes a blue shift on one side of the string only. This generates a step-function increase of temperature $\Delta T/T \sim 10^{-3}$ on the angular scale of the horizon at recombination ~ 1 deg (Kaiser & Stebbins, 1984).

(d) Primordial fluctuations

In the absence of any legacies from the GUT (Grand Unified Theories) era there must be a modulation due to primordial fluctuations in which density variations are expected to have a power law spectrum varying as (wavenumber)$^{-1}$. A variety of cosmological models predict adiabatic perturbations of this form. The associated temperature fluctuations increase with angular size, up to the scale of the horizon at recombination, as shown in Fig. 3.2. Observations have already ruled out some models but in a flat universe dominated by dark matter a detection sensitivity exceeding $\Delta T/T \sim 10^{-6}$ is required to draw strong cosmological deductions. It is evident that the ability to map fluctuations ΔT on angular scales from a few arcmin to 1 deg would be of great interest provided that the required sensitivity can be achieved. A glance at the world's instrumentation shows that no suitable radio telescope exists to perform the required observations; the angular resolution of a synthesis array is limited by the smallest available baseline and this is always too large. We are currently enhancing the 5-km telescope

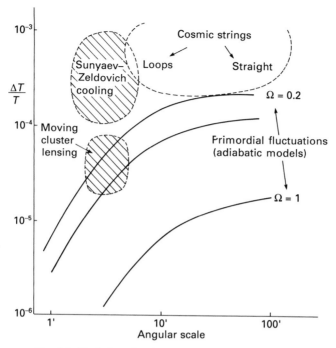

Fig. 3.2. Variations in the microwave background due to primordial fluctuations in comparison to other possible sources.

at Cambridge so that the array can be operated in a densely-packed mode with cooled receivers and a bandwidth of 350 MHz. This will achieve in one month's observing time a sensitivity $\Delta T/T \sim 2 \times 10^{-5}$ at 5 GHz with a resolution of 6 arcmin. To map the largest scales requires a densely packed array on baselines up to about 10 m. A suitable name for this instrument would be the VSA (very small array).

Although a sensitivity near $\Delta T/T \sim 10^{-5}$ has been achieved using single dishes in a beam-switched mode these are advantages in using arrays.

Problems of spillover, man-made interference and standing waves in the feed are minimized when antennas are correlated, and mapping is straightforward. An array is also adaptable because the synthesized beam can be matched to the source. For a thermal source the detection sensitivity is greatest when the source just fills the beam, and a larger effective aperture only increases the confusing signal from unresolved non-thermal sources. The steeper spectrum of non-thermal radiation favours the use of high frequencies, but noise from the troposphere is then increased and the best compromise is obtained near 15 GHz.

A suitable system for mapping structure in the microwave background

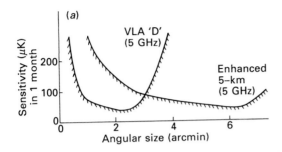

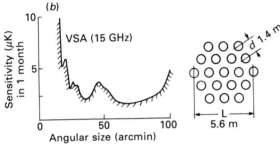

Fig. 3.3. (*a*) Comparison of the sensitivities of the VLA and enhanced 5-km telescope as a function of angular scale (observing time = 1 month). (*b*) Projected sensitivity of the VSA (observing time = 1 month) as a function of angular scale for the array configuration shown on the right (*d* = twice horn diameter = 1.4 m).

would be a hexagonal array containing 20 dishes 0.7 m in diameter on baselines up to 10 m. For a bandwidth of 600 MHz and a system temperature of 50 K a sensitivity of the order $\Delta T/T \sim 10^{-6}$ would be achieved in one month over angular scales in the range 20 arcmin to 1 deg (see Fig. 3.3).

There are, of course, some potential difficulties such as small-scale turbulence in the troposphere, thermal and non-thermal sources in our Galaxy, and the possible existence of numerous faint radio galaxies at large redshifts. All of these need very careful consideration but on the available evidence no single problem is insuperable. Non-thermal components from the galaxy are likely to be the most serious obstacle but they can be removed by observations with linearly scaled arrays at 5, 10 and 15 GHz. Unresolved confusing sources can be mapped at much higher resolution with the 5-km telescope and subtracted. Genuine cosmological signals are expected to be linearly polarized so the VSA must also be endowed with polarization capability.

We conclude that the challenging new field of mapping small-scale structure in the microwave background presents rich opportunities for exciting advances in observational cosmology.

References

Birkinshaw, M., Gull, S.F. & Hardebeck, H. (1984). *Nature*, **309**, 34.
Birkinshaw, M. & Gull, S.F. (1983). *Nature*, **302**, 315.
Chase, S.T. (1986). *Nature*, **323**, 42.
Kaiser, N. & Stebbins, A. (1984). *Nature*, **310**, 391.

4

VLBI from ground and space

R. T. SCHILIZZI

Summary

Instrumental progress in astronomy has been driven by three principal motives: to improve sensitivity, spectral resolution, and angular resolution. The most spectacular advance in angular resolution has come in the radio using the very-long-baseline interferometry (VLBI) technique, which for Earth-based systems provides almost 10^5 times more resolving power than the largest fully steerable single radio telescope. This article attempts to describe the influence of VLBI on astronomy by sketching some of the scientific achievements of VLBI thus far, and then considering aspects of Space VLBI. Current instrumental subsystems in VLBI and expected developments are not described; an excellent review is given in Chapter 9 of Thompson, Moran & Swenson (1986).

Introduction

Twenty years ago, the first VLBI experiments were carried out with two-element tape-recording interferometers (Carr *et al.*, 1965; Broten *et al.*, 1967; Bare *et al.*, 1967). Two years ago, a 17-element global VLBI array was used to image complex sources on milliarcsecond to arcsecond scales. Two months ago, the first fringes were detected on baselines from Earth to an antenna in space (Levy *et al.*, 1986), heralding in the age of Space VLBI.

In this 20-year period the VLBI technique has gone through a substantial transformation, a transformation that is still continuing. Whereas 20 years ago, detection of fringes was the state-of-the-art, the thrust at the present time is towards high image quality and towards yet higher angular

23

resolution. This is necessary to satisfy astronomers' demands to measure details in structures of compact radio sources in a wide variety of objects ranging from stars to the active nuclei of galaxies and quasars.

High image quality requires good aperture (u–v) plane coverage and accurate calibration, both of which are prime goals of ground-based developments now underway or planned. In the USA, funding has been secured for a ten-element very long baseline array (VLBA) which should be completed in 1992 (Kellermann & Thompson, 1985). This will be the first dedicated VLBI system in the world, providing not only high quality images but also full-time operation, including the possibility for observations of 'targets of opportunity' which the part-time networks of the present day find hard to accommodate. Joint observations together with the Very-Large-Array (VLA) will ensure excellent u–v coverage down to short inter-ferometer spacings. In Europe a dedicated VLBI array is a project that must await a greater degree of political cohesion than is presently the case. However, the current European VLBI Network (EVN) has two unique features which will ensure its place beside the VLBA in the arrays of the future: large collecting area telescopes for very sensitive observations, and the possibility of simultaneous observation with the MERLIN array in the UK to provide essentially uniform u–v coverage from near-zero spacing out to about 2000 km. Additional spacings, out to the maximum possible on Earth, are generated by cooperative observations with US telescope primarily, but in the future could also include baselines to the 70 m Suffa telescope near Samarkand in the USSR. Comparable aperture plane coverage to the VLBA can be achieved, particularly for northern sources. What does appear to be within the realms of financial possibility in Europe is European funding of a large data-processing facility to augement the EVN in the 1990s. A proposal to this effect in under consideration.

High angular resolution can be achieved through the use of shorter wavelengths for Earth-based observations (mm-VLBI) and through exten-sion of the baselines by an element in space. The concept of Space VLBI is one which had its origins in the early days of VLBI, but which only in the last three or four years has begun to appear an achievable goal with the funding of the Soviet RADIOASTRON mission and the detailed study of the QUASAT mission in the West. Space VLBI has the additional advantage that the aperture plane coverage provided by the orbiting element together with the global array is substantially better than for the global array alone. The combination of angular resolution and image quality resulting is, in fact, unattainable with ground-based arrays alone.

Table 4.1 *VLBI milestones*

1965	First tape-recording interferometry on Jupiter (Carr *et al.*, 1965)
1967	First (V)LBI on QSO'S and galaxies (Broten *et al.*, 1967; Bare *et al.*, 1967)
	First VLBI on cosmic masers (Moran *et al.*, 1967)
1970	First discussions on Space VLBI (Burke, 1984)
1971	First detection of super-luminal motion (Whitney *et al.*, 1971)
1972	First multi-station observations in USA (Cohen *et al.*, 1975)
1975	US VLBI Network established
1976	First multi-station observations in Europe (Schilizzi *et al.*, 1979)
	Hybrid mapping technique first developed (Readhead & Wilkinson, 1978)
1980	European VLBI network established
1981	First millimetre-wavelength VLBI (Readhead *et al.*, 1983)
1982	QUASAT proposed to ESA and NASA
1985	VLBA funded in USA
	RADIOASTRON funded in USSR
	Venus balloons tracked with VLBI (Sagdeev *et al.*, 1986)
1986	QUASAT approved for Phase A in Europe
	TDRSS Space VLBI demonstration (Levy *et al.*, 1986)
1989?	Long-baseline-array of Australia Telescope in operation
1991?	RADIOASTRON launch
1991–2?	European VLBI data processing facility in operation?
1992–3?	VLBA in full operation
1995?	QUASAT launch?

Table 4.1 lists a number of scientific and organisational milestones of VLBI in the last two decades, as well as a look ahead to future milestones.

VLBI–the present

Typical observations now involve eight telescopes in a European–US network (the so-called global network), a number which is kept down by a combination of shortage of tape for wide bandwidth observations, and small-capacity data processors which necessitate multiple passes of the data tapes to make all pair-wise correlations. Despite these restrictions, some fascinating results have emerged over the years. I will discuss a number of these in order to illustrate the use of the VLBI technique in astronomy.

3C345, the archetypal super-luminal source

Many of the new VLBI results have come about in the context of proper motion studies. The best-known of these is super-luminal motion, and the most carefully studied of the super-luminal sources is the quasar 3C345. Fig. 4.1 is derived from 22GHz observations by Biretta, Moore & Cohen (1986) and shows a radiating blob moving to the west, and rotating in position angle as it moves. Astrometric measurements by Bartel *et al.* (1986) of 3C345 with respect to the nearby quasar NRAO512 have shown that the eastern component of 3C345 has no proper motion relative to NRAO512 and can therefore be assumed to be the stationary core component coincident with the nucleus of the quasar. It is the western component that moves super-luminally. The emission is clearly on one side of the core, a feature seen in all super-luminal sources, and it appears to be a succession of blobs with no

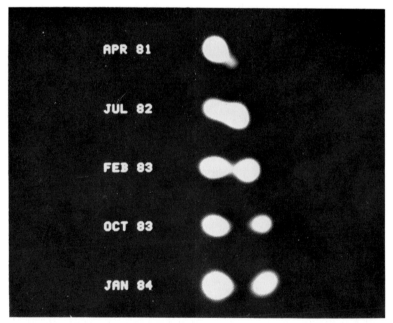

Fig. 4.1. Superluminal motion in 3C345. VLBI images of the quasar 3C345 ($z = 0.595$) at 22.2 GHz at 5 epochs (Biretta *et al.*, 1986). The component separation at Jan 84 is 0.84 milliarcsec or 3.2 pc ($H_0 = 100$ km sec.$^{-1}$ Mpc^{-1}; $q_0 = 0.5$). At each epoch the left-hand component is the 'core' and the new component is at the right. This new component changed position angle and apparently accelerated from a speed of three times that of light to a speed of 5.9 times that of light as it moved away from the core. It also caused the largest flux outburst ever seen in this object between 7 and 300 GHz.

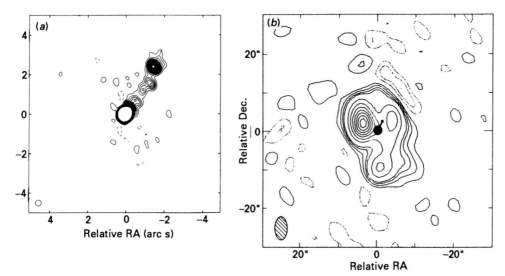

Fig. 4.2. Large-scale structure of 3C345. (*a*) MERLIN map of 3C345 at 1666 MHz (epoch 1980.9) (Browne *et al.*, 1982). Contours are 0.1, 0.2, 0.3, 0.4,... % of the peak brightness of 6.26 Jy per beam. Beam size is 0.25 × 0.25 arcsec. (*b*) 5 GHz map of 3C345 made with the Westerbork Synthesis Radio Telescope in redundancy mode (Schilizzi & de Bruyn, 1983). The core component (10.15 Jy) and the arcsec jet (0.35 Jy) have been removed from the map as indicated. Contours are − 0.025, − 0.0125, 0.0125, 0.025, 0.0375, 0.05, 0.1, 0.15,..., 0.4% of the core flux density.

underlying continuous stream of emission. The projection of the trajectory of the western blobs backwards towards the nucleus shows that it does not intersect the nucleus. It is not clear whether this indicates that the motion occurs along a curved path or whether the true nucleus is not visible and the motion is ballistic (Cohen *et al.*, 1983). The acceleration seen in the motion of the western component between April 1981 and February 1983 might be a consequence of motion along a curved path.

The popular explanation for super-luminal motion is based on relativistic bulk motion of the emitting material towards the observer close to the line of sight. It is a derivative of a theory put forward by Rees (1967) to explain flux density variations in quasars. Relativistic beaming can account for the motion itself, the one-sideness of the emission as an effect of aberration, and the curvature of the path as a magnification of a small intrinsic curvature.

However, in many of the 20 or so (Porcas, 1987) super-luminal objects large-scale structure is observed (see Fig. 4.2). If the apparent dimensions of this structure are de-projected by the angle to the line of sight given by the relativistic beaming model, these objects are all the largest or amongst the largest in the universe at their particular redshifts, which seems unlikely for objects thought to be selected at random (Schilizzi & de Bruyn, 1983). This difficulty can be avoided if the axis of ejection were to precess over cosmological time scales or if bending occurs in the ejection path on the large scale. It is not so easy to circumnavigate cases where the nuclear structure is extremely well aligned with the extended structure (e.g. 1721 + 34, Barthel *et al.*, 1985*a,b*, 1987).

We may have to resort to further modifications of relativistic beaming such as sideways moving shocks in the line of sight occurring in bulk relativistic flow at large angles to the line of sight (Blandford & Königl, 1979). Perhaps we should also re-investigate light echo effects from screens (Miley, 1983; Norman & Miley, 1984; de Waard, 1986). The final word is not in on super-luminal sources and on whether they are an important element in our understanding of energy flows in active sources.

Polarization VLBI

One of the most important recent developments in VLBI is the determination of the polarization structure of selected compact objects (including 3C345. Fig. 4.3) by Wardle *et al.* (1986). The diagram shows that the opaque core to the east is nearly unpolarized, and the western extension is strongly ($\leqslant 1\%$) polarized with the magnetic field rotating from alignment with the outflow direction to perpendicular on angular scales of a few milliarcsec (a few tens of parsec). The change in magnetic field direction is expected on simple models of adiabatically expanding jets. These are technically difficult measurements to make at present due to the fact that polarized emission is weak and that instrumental polarization effects at each telescope differ substantially from each other because no two telescopes are alike. The VLBA promises to be a good polarimeter since all telescopes are identical. Polarization VLBI is an area of future research with great promise as it provides the radio astronomer with his most powerful diagnostic tool for the properties of the plasma in active nuclei.

3C119 – a compact steep spectrum source

Compact steep spectrum sources are now recognized as a third class of radio source intermediate between the classical steep spectrum doubles

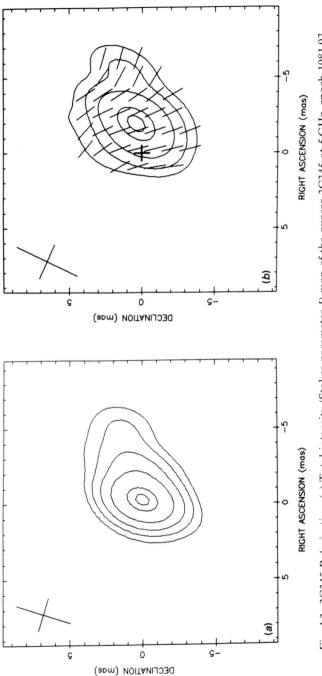

Fig. 4.3. 3C345 Polarization. (*a*) Total intensity (Stokes parameter *I*) map of the quasar 3C345 at 5 GHz, epoch 1981.93 (Wardle *et al.*, 1986). The contour levels are −5, 5, 10, 20, 40, 80, and 95% of the peak brightness of 6.4 Jy per beam. The restoring beam is shown by the cross in the corner and has dimensions (FWHM) of 3.9 × 2.3 millarcsec in position angle −18 deg. (*b*) Polarized flux density, $P = (Q^2 + U^2)^{1/2}$, map of 3C345 at 5 GHz (Wardle *et al.*, 1986). The contours are 10, 20, 40, 80, and 95% of the peak brightness of 194 mJy per beam. The location of the compact core is indicated by the cross. The restoring beam has dimensions of 4.7 × 2.5 milliarcsec in position angle −25 deg.

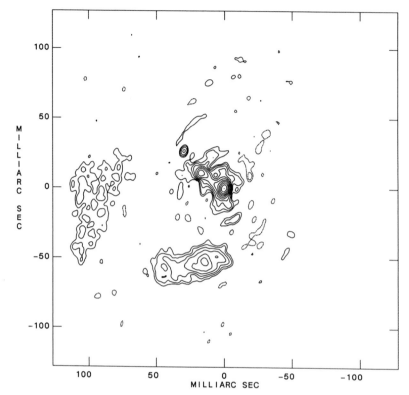

Fig. 4.4. 3C119. 1.66 GHz image from combined ten-station VLBI network and five-station MERLIN observations (van Breugel *et al.*, 1987). The nucleus of the quasar is the weak compact component to the north-east of the brightest structure. Contours are -0.2, -0.1, 0.1, 0.2, 0.4, 0.8, 1.6, 3.2, 6.4, 12.8, 25.6, 51.2, 98% of the peak flux density (2.50 Jy). The restoring beam is 7×5 milliarcsec in position angle -10 deg.

and the compact flat spectrum sources. Their identifying characteristics are physical sizes of the order of a few kpc and no dominant flat spectrum core. 3C119 is an example of this class. A combination of a ten-station VLBI and a five-station MERLIN observation at 1.66 GHz (van Breugel *et al.*, 1987) shows a swirling structure (Fig. 4.4) buried deep in the nucleus of this 20^m quasar.

Fig. 4.4 is 250 milliarcsec on each side, equivalent to 0.9 kpc if the redshift is 0.408 (Fanti *et al.*, 1986). Observations at 6 cm with similar resolution show that the weak compact component to the north of the structure in the centre of the diagram has an inverted spectrum, and is therefore likely to be the true nucleus (Fanti *et al.*, 1986). We may ask whether this is a jet associated with a slow precession of the nucleus or a jet carried around by rotating gas in the

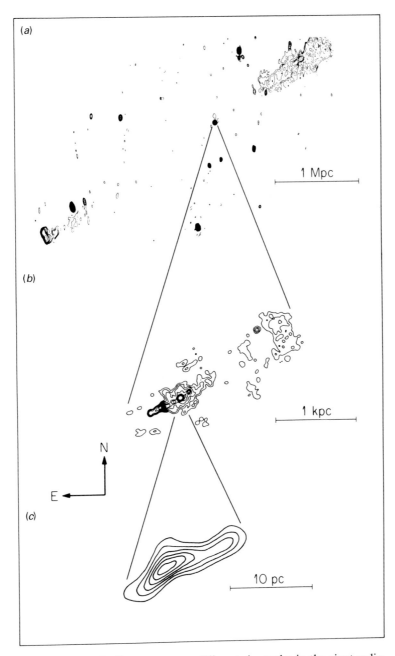

Fig. 4.5. 3C236. Radio structure on different size scales in the giant radio galaxy 3C236 ($z = 0.098$). (*a*) A 1.4 GHz image made with the Westerbork Synthesis Radio Telescope; (*b*) a 1.66 GHz image derived from a combined VLBI and Jodrell Bank MERLIN measurement; (*c*) a 5 GHz VLBI image (Barthel *et al.*, 1985).

line emitting region of the quasar? Optical observations of line velocities would be very helpful in this respect but difficult to carry out. The jet appears on only one side of the nucleus, in common with super-luminal sources. For a variety of reasons it is unlikely that Doppler boosting plays a significant role in 3C119; the one-sideness may, in this case, be intrinsic.

Fig. 4.4 is also interesting from a technical point of view. It is a high quality image by current VLBI standards with a dynamic range (peak flux density to noise in the image) of between 1000 and 1500. Examination of the noise distribution over the image reveals systematic patterns at a dynamic range of 2000 to 1, the origin of which is not yet understood. It does indicate that

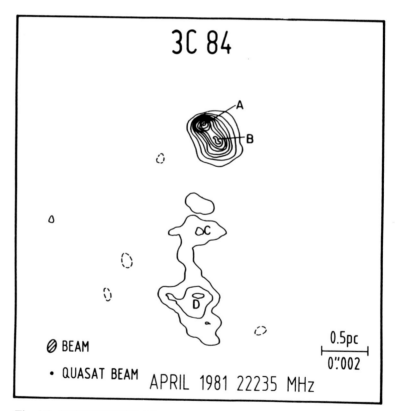

Fig. 4.6. 3C84/NGC1275. The radio structure of the relatively nearby galaxy NGC1275 at 22 GHz (Readhead *et al.*, 1983). The structure in all regions is variable on a time scale of a half year. The size of the emission region shown here is about 100×25 QUASAT beams. The resolution provided by QUASAT will enable us to probe the structure on a scale of 10^{17} cm, i.e. comparable to the expected accretion disk around a $10^9 M_\odot$ black hole.

high-quality images are possible in VLBI, just as they are in connected element interferometry. One can expect images of this or better quality to be routine in VLBI in the future.

The giant radio galaxy 3C236

My third example is the nucleus of the radio galaxy 3C236 (Fig. 4.5). Panels (*b*) and (*c*) show the small-scale structure is almost precisely aligned with the very large-scale lobes (panel (*a*)), indicating that the mechanism producing the small-scale jet has an exceptionally long memory for direction and is able to collimate the jet very effectively on scales of parsecs or smaller. A massive black hole is an obvious candidate. Note how the radio emission on the small scale is only seen out to the kpc range, even though we know from the spectrum of the outer lobes that energy must be being supplied to them along invisible jets. It may well be that radio-emission is only generated when there is interaction with thermal material in the path of the jet. There is similar evidence for this suggestion in a number of objects at smaller distances than 3C236, e.g. 3C277.3 (van Breugel, *et al.*, 1985).

NGC1275

Example number four is the well-known radio source in the dominant galaxy in the Perseus cluster – 3C84/NGC1275 (Fig. 4.6). Here we see a 22-GHz image by Readhead *et al.* (1983) with an interferometer beamsize of about 3×10^{17} cm. This image clearly established the core-jet nature of the source which had not been obvious in earlier lower resolution observations, and showed that the core itself was elongated perpendicular to jet axis. It is not implausible in a massive galaxy like NGC1275 that this elongation represents radiation from an accretion disk of dimension a few times 10^{17} cm. More recent observations at 22 GHz by D.C. Backer & Readhead (private communication) have shown remarkable changes in the position angle of the structure in the core on timescales of months. They speculate that there may be an SS433-type jet involved. These measurements are at the limit of resolution. An imaging device with longer baselines seems called for. Just such as instrument is QUASAT, as can be seen from the QUASAT beam shown on the diagram.

M82

My next example is the nearby galaxy M82. Recent European observations at 5 GHz by Wilkinson & de Bruyn (1987) appear to have resolved an old controversy. M82 is an irregular galaxy in the optical, it is a strong IR source, and has at least 20 compact radio sources in its disk of which the strongest lies some 10 arcsec away from the kinematic centre of the galaxy. The nature of this bright source has always been controversial, the two contending hypotheses being a classical active nucleus or a supernova remnant. Lower resolution VLBI at 18 cm showed an elongated shape (Fig. 4.7), almost centrally peaked, and consistent with a core with a jet either side, but the 6 cm data seen in Fig. 4.8 show conclusively that the structure is shell-like and that the source is very likely a supernova remnant.

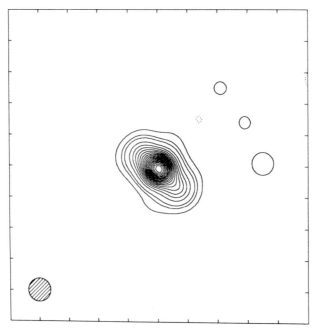

Fig. 4.7. 1.66 GHz VLBI map of the compact radio structure of the source 41.9 + 58 in M82 (Wilkinson & de Bruyn, 1984). Peak flux density is 136 mJy beam^{-1}. Beamsize is 15 milliarcsec, and the axes are graduated in units of 20 milliarcsec ($\sim$ 320 light days).

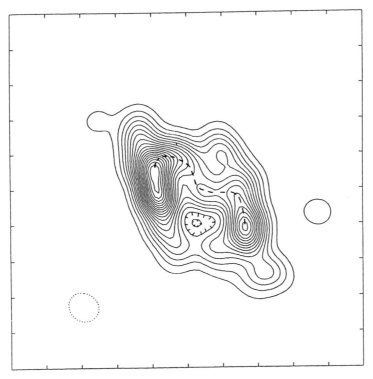

Fig. 4.8. M82. The 5 GHz compact radio structure of the source 41.9 + 58 in M82 (Wilkinson & de Bruyn, 1987). A line has been drawn along the ridge to draw attention to the shell-like structure which suggests that it is, in fact, a supernova remnant and not a classic active nucleus. The axes are graduated in units of 6 milliarcsec (100 light days). The beamsize is 5 × 5 milliarcsec. Peak flux density in 9 mJy.

Distance to our galactic centre

My last two examples are to be found in our own galaxy, and both are beautiful illustrations of the power of proper motion studies. Water masers move around in the sky, sometimes with a systematic motion when they are in an outflow region near a newly formed star, sometimes with a random motion. Application of the well-known optical techniques of cluster and statistical parallax to the H_2O maser motions can be made by taking the ratio of the line-of-sight Doppler shifts of the maser components and their angular proper motions and assuming a model for the motion to give a distance. In the case of Sgr B2(N) (Fig. 4.9) which lies within 0.3 kpc of the galactic centre, the distance has recently been established as 7.1 ± 1.2 kpc by

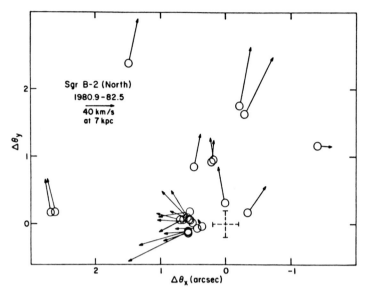

Fig. 4.9. Sgr B2(N). Proper motions of H_2O maser spots in Sgr B2 (north). The position of each maser spot is indicated by the centre of each circle (the maser spots are much smaller than the circles), and the direction and magnitude of its proper motion by the arrows. The masers appear to be expanding from a central source (presumably a massive newly formed star) whose position is indicated by the dashed error bars (Reid et al., 1986).

H_2O maser proper motion measurements by Reid et al. (1986) a substantial decrease in the value assumed for the last decade or more.

Application of this technique to H_2O masers in nearby galaxies can be made either by statistical parallax or orbital parallax (Reid, 1984; Reid et al., 1987) to give a completely independent measure of the distance to these objects and thus, by extrapolation, the scale of the universe and the parameter of that scale, the Hubble Constant. This will be one of the prime goals of QUASAT.

Fig. 4.10. SS433. (a) 5 GHz maps of SS433 made at intervals of two days, starting on 17 May 1985 (Vermeulen et al., 1987). The mean Julian Dates (− 2440000) of the observations are: 2603.643, 2605.629, 6207.623, 6209.630, 6211.628, and 6213.597. Contour levels have been drawn at − 2, 2, 4, 8, 16, 32, 64 and 90% of the peak flux level in the map. The ellipse at lower left indicates the FWHM of the restoring beam. The curve in each map represents the locus of ejecta as predicted by the kinematic model parameters of Anderson et al. (1983); the markers are drawn at intervals of two days along the curve. (b) Crosscuts through the maps of (a), along the kinematic model curve shown. The horizontal scale is labelled in time units (days) since the assumed ejection. All cross-cuts have the same flux density scale, enabling one to follow the flux behaviour of features in the map from day to day.

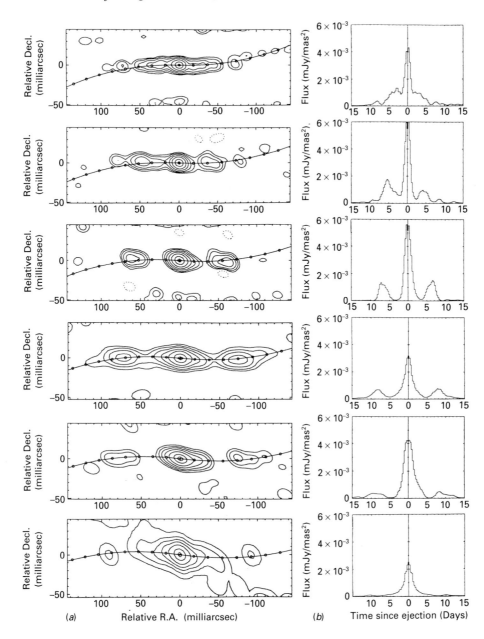

(a)   Relative R.A. (milliarcsec)   (b)   Time since ejection (Days)

SS433

SS433 is a galactic X-ray binary star with oppositely directed jets of gas moving at a speed of 0.26c along the axes of a precessing accretion disk surrounding the compact secondary. VLBI observations of the radio emission have sufficient resolution $(8 \times 10^{14}\,\text{cm}, 10$ milliarcsec) at the distance of SS433 (5.0 kpc) that the motions of individual ejecta along ballistic paths can be directly measured.

Fig. 4.10(a) shows a remarkable set of VLBI images made by Vermeulen *et al.*(1987) with the European network at 5 GHz at two-day intervals in May 1985. The radio emission is seen to occur in discrete blobs rather than in a continuous jet (as is thought to be the case in the optical as well) and to move outwards as a function of time. The maps are quite symmetrical if we assume the brightest knot is coincident with stellar source, and the locus of the radio emission conforms well to the predictions of the kinematic precessing beam model (Abell & Margon 1979).

The apparent proper motions of the peaks on either side of the central feature are, however, considerably smaller than predicted. This discrepancy has been resolved by determining that the blobs in each map can be decomposed into knots or 'bullets' which do in fact move at 0.26c. Fig. 4.10(b) shows that the intensity of these bullets increases at an angular distance of about 50 milliarcsec from the central feature (a clear example occurs in the west beam between epochs 2 and 3) and then decays rapidly, probably due to adiabatic expansion. A possible explanation for this 'brightening zone' is that it represents the region where a new blob or bullet, travelling in a different direction to the previous blob due to precession, pierces the bowshock of its predecessor causing particle acceleration and/or magnetic field compression, and thus enhanced emission.

VLBI—the instrumental future

Where is VLBI going in the future? In Table 4.1, I named a number of the milestones ahead: the VLBA, a solid milestone; the European VLBI data processing facility, an accreting milestone; RADIOASTRON, a solid milestone; QUASAT, another accreting milestone. We can also expect continuing improvements in image reconstruction techniques (see Readhead, Chapter 7, these proceedings).

Centimetre-wavelength VLBI

The two major elements in centimetre-wavelength VLBI in the 1990s will be the US VLBA (Kellermann & Thompson, 1985) and the European VLBI network. Fig. 4.11 shows the locations of the radio telescopes in these two arrays, as well the USSR VLBI network and other telescopes expected to have substantial observing time available for VLBI. In the 1990s a standard observation is likely to make use of 15 or more telescopes to enhance the aperture plane coverage and thereby improve the image quality. The data-processing facilities planned for the VLBA, and proposed for the EVN, will have sufficient capacity to enable correlation of the tapes from all these elements in one pass. Another important aspect of the processor capacity is significantly increased spectral capability for studies of maser sources with 256 to 512 spectral points per interferometer baseline.

Millimetre-wavelength VLBI

There has been a spate of building millimetre-wave telescopes in recent years in a number of areas of the world. It is now feasible to think of establishing ad-hoc networks of telescopes operating at 7 and 3.6 mm wavelength to carry out VLBI on strong sources with increased resolution over that possible on Earth at centimetre wavelengths.

Space VLBI

Certainly, one of the most exciting concepts for the future is Space VLBI. For many astronomical problems, these is a clear need to obtain uniform aperture plane coverage to produce much higher quality images, while at the same time extending the baselines out as far as is possible. Ground-based millimetre-wave VLBI will certainly have its place in providing higher angular resolution, but it will not provide the aperture plane coverage required for high-quality images because there are too few large antennas at these wavelengths. A VLBI element in space can, however, achieve both goals of high image quality and increased angular resolution.

The technical feasibility of Space VLBI has been succinctly demonstrated by Levy *et al.* (1986) in their successful experiment to combine a 5 m diameter

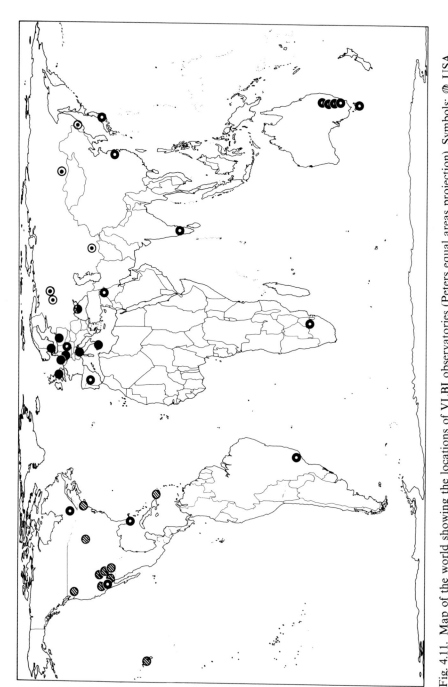

Fig. 4.11. Map of the world showing the locations of VLBI observatories (Peters equal areas projection). Symbols: ⦿ USA VLBA; ⊙ USSR Network; ● EVN; ○ other telescopes.

antenna on the geostationary satellite TDRSS 1 (tracking and data relay satellite system) with 64 m diameter antennas in Japan and in Australia. An interferometer spacing of 1.4 Earth diameters was created in their first experiment and all three southern radio sources observed showed fringes.

Four space agencies are, or have been, studying Space VLBI concepts. The Space Research Institute/Interkosmos in the USSR has taken the lead by approving the RADIOASTRON programme for two radio telescopes in space (see Schilizzi 1987*a*). ESA and NASA spent two years (1983–85) in assessment studies of a joint project, QUASAT (see ESA documents SCI (85)5 and SP-213, Schilizzi *et al.*, 1984; Schilizzi, 1987*b*). However, the NASA Explorer Program has recently suspended evaluation of potential space astronomy missions, including QUASAT, as a result of budget problems arising from the Challenger crash. ESA is currently investigating other mechanisms of collaboration before continuing with a phase A study on QUASAT approved earli in 1986. The Japanese Institute of Space and Astronautical Science Science (ISAS) is the fourth agency carrying out a study of Space VLBI.

A schematic diagram of the space/ground VLBI system is shown in Fig. 4.12. The space-borne antenna will be capable of observing in both hands of circular polarization simultaneously for any of the wavelength complement (QUASAT will have the extra possibility of simultaneous dual frequency, dual polarization reception), and will relay the received signals via a digital or analogue link directly to telemetry stations on the ground. A

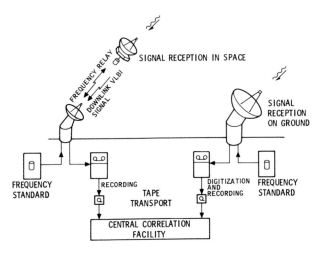

Fig. 4.12. Schematic diagram of a space/ground VLBI system.

phase/frequency references for the antenna in space, stable to about 1×10^{-14}, will be based on hydrogen maser oscillators on the ground and relayed directly to the satellite via a two-way link from the telemetry stations in turn. All 'house-keeping' communication with the space element will be through one or more telemetry stations in the network. The parameters of QUASAT are listed in Table 4.2.

After transmission to the ground, the signal will be recorded on magnetic tape in digital form, and transported to the central processing facility of the European or US VLBI array for correlation with similar tapes from the ground VLBI arrays. After correlation and calibration the data will be sent to the principal investigators for further analysis.

The antenna concepts for RADIOASTRON and QUASAT are shown in Figs. 4.13 and 4.14. The central 3 m of the RADIOASTRON antenna is solid,

Table 4.2 *Parameters of QUASAT*

orbit	
apogee altitude	12 500 km
perigee altitude	5700 km
inclination	63.4 deg
period	5.3 h
angular resolution	75×10^{-6} arcsec at 22 GHz
imaging capability	excellent
antenna diameter (m)	15
antenna type	inflatable centre-fed
f/D ratio	0.42
feed configuration	coaxial at prime focus
payload mass	950 kg
observing frequencies (GHz)	22, 5, 1.6, 0.3
receiver technology	HEMT (high electron mobility transistor)
receiver cooling technology	Stirling cycle
polarization	dual
datalink bandwidth per polarization (MHz)	32
rms map noise (48ʰ observations)	$\leqslant 1$ mJy (22 GHz)
	$\leqslant 0.5$ (5 GHz)
	$\leqslant 0.3$ (1.6 GHz)
	$\leqslant 3$ (0.3 GHz)
status	ESA Phase A Study approved February 1986
launch	1994–95
lifetime	
design	2 yr
operational (expected)	5 yr

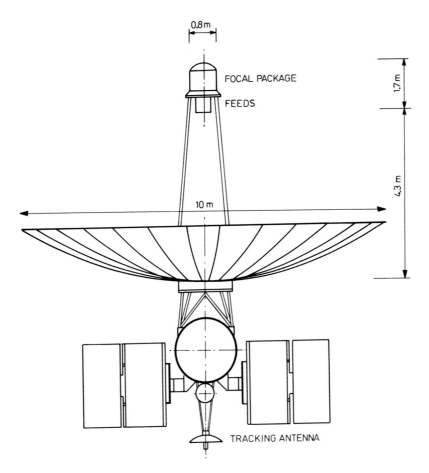

Fig. 4.13. Concept for RADIOASTRON antenna.

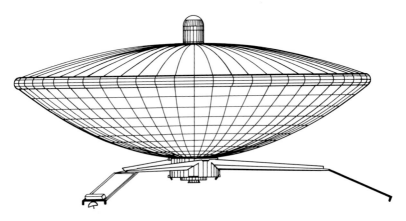

Fig. 4.14. ESA concept for QUASAT antenna.

providing the hub around which the deployable petals are hinged. During launch the petals are folded upwards around the focal box which is itself supported on three legs. After the spacecraft reaches the correct orbit the petals fold outwards to create the 10 m aperture.

The ESA concept involves new technology – inflatable structures which can be rigidized in space. The antenna is launched folded and, in space, gas is blown into the balloon to inflate it to the required dimensions. The KEVLAR-based material in the balloon is pre-cut to the correct form and cured rigid in orbit by the action of the sun on a resin impregnated into the material. After rigidization the gas is vented into space.

The frequencies of observation chosen for QUASAT, and thereafter by RADIOASTRON, are 22, 5, 1.6 and 0.3 GHz. The maser line emissions at 22 GHz for H_2O and 1.6 to 1.7 GHz for OH dictate two of the frequency complement. 22 GHz is at the same time the highest operating frequency of the ground network for which there are telescopes of large collecting area, and for which good sensitivity for both continuum and line can be achieved with modest diameter antennas in space. 5 GHz is approximately the geometric mean of the other two, and provides a different combination of resolution and surface brightness sensitivity. The lowest frequency, 0.3 GHz, provides the capability to investigate refractive scattering as a possible cause of low frequency variables as well as to observe pulsars.

RADIOASTRON and QUASAT are complementary in aim. The principal difference lies in the orbits chosen–RADIOASTRON aims to obtain crude structural information at very high angular resolution; QUASAT aims to obtain high-quality images at less extreme resolutions, and with higher sensitivity.

The orbits of QUASAT and RADIOASTRON 1 have been the subject of considerable study. The RADIOASTRON 1 orbit will achieve angular resolutions up to ten times that of the ground-based global arrays, but with rather crude imaging capability. On the other hand, the QUASAT orbit is presently optimized for near-perfect aperture plane coverage and, consequently, very high-quality images. The angular resolution is a factor of three higher than for the ground-based arrays, which corresponds to a factor of ten smaller area for a resolution element in the resulting images.

The impact of this imaging capability can be seen in Fig. 4.15 where a simulated radio source (panel *a*) is observed with the European VLBI network augmented by east- and west-coast US telescopes (panel *b*), and with the same ground network together with QUASAT (panel *c*).

Two additional advantages of the QUASAT orbit are the ability to make

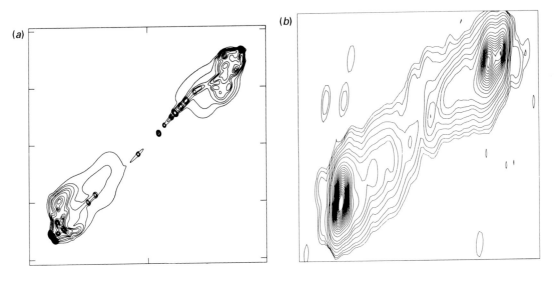

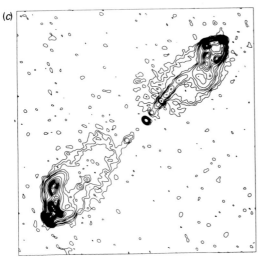

Fig. 4.15. Image simulation with QUASAT. (*a*) Model brightness distribution. The largest angular size is ~ 10 milliarcsec. (*b*) Simulated image of the model at declination 0 deg using the agumented EVN. (*c*) Simulated image of the model at declination 0 deg using the augmented EVN + QUASAT. (Diagrams are due to P.N. Wilkinson in ESA SCI (85)5.)

relatively crude images on short timescales (3 hours), which will be useful in following structural evolution in variable galactic stellar sources, and the ability to make images in the southern hemisphere where ground radio telescopes are rather scarce.

Table 4.3. *Spatial resolution of QUASAT at 22 GHz*

Object	Distance[a]	Linear size[b]
Galactic centre	7.1 Kpc	1.0×10^{13} cm, 0.7 astronomical units
Centaurus A	5 Mpc	5.7×10^{15} cm, 2.2 light days
NGC1275 (3C84)	53 Mpc	5.8×10^{16} cm, 0.75 light months
3C273 ($z = 0.158$)	491 Mpc	4.2×10^{17} cm, 5 light months
3C345 ($z = 0.595$)	1992 Mpc	9.2×10^{17} cm, 1 light year

Notes:
[a] $H_0 = 100 \, \mathrm{km \, s^{-1} \, Mpc^{-1}}$, $q_0 = 0.5$.
[b] Corresponding to an angular resolution of 75 microarcsec.

The linear size of the QUASAT resolution element at 22 GHz is given in Table 4.3 for a variety of benchmark objects. The equivalent values for RADIOASTRON 1 are three times smaller. The resolution is, broadly speaking, $\leqslant 1$ AU for galactic objects, 0.01–0.1 pc for nearby galaxies, and ~ 1 pc for distant radio galaxies and quasars.

International cooperation

Both RADIOASTRON and QUASAT are international projects. Plans exist for space hardware participation by western institutes in RADIOASTRON, and for world-wide participation in QUASAT, including, hopefully, a substantial US component when the Explorer programme gets underway again. The availability of the global complement of radiotelescopes is, of course, an essential element of the Space VLBI concept.

The launch of the second RADIOASTRON satellite could be adjusted to coincide with that of QUASAT if the latter actually does fly. The orbit of RADIOASTRON 2 would then be tailored to QUASAT's to provide continuous aperture plane coverage out to 55 000 km, and high-quality imaging with resolution elements of 40 microarcsec diameter at 22 GHz. Fig. 4.16 shows the u–v coverage for QUASAT + RADIOASTRON 2. In addition, Japanese scientists are assessing the feasibility of a 10 m diameter antenna in Earth orbit, which could be launched in the 1995 time frame to create a multispacecraft VLBI system together with QUASAT and RADIOASTRON 2.

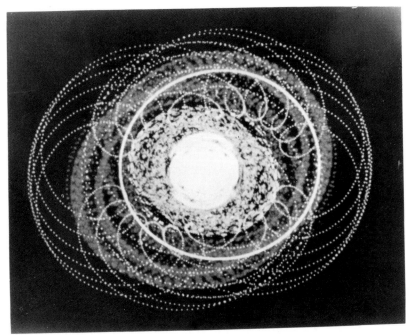

Fig. 4.16. Aperture plane coverage for a space/ground VLBI system consisting of two space elements, QUASAT and RADIOASTRON 2, in adjacent orbits (maximum altitudes above the Earth's surface 12 500 and 30 000 km) and a global VLBI array on the Earth. The longest baselines generated are about 55 000 km. A number of regions can be identified in this diagram. The innermost area is the aperture plane coverage from the ground array alone; the elliptical annulus adjacent to this is the aperture plane coverage generated by QUASAT and the ground network; the grey region beyond that is the coverage from RADIOASTRON 2 and the ground network; and the dotted curve results from the spacecraft-to-spacecraft interferometer. The white circle shows, for comparison, the orbit of RADIOASTRON 2; the size the of the Earth is approximately that of the innermost part of the diagram.

Acknowledgements

I thank Mr Nan Rendong, Dr P.N. Wilkinson and Dr J.F. Jordan for supplying Figs. 4.4, 4.8 and 4.16 respectively.

References

Abell, G.O. & Margon, B. (1979). *Nature*, **279**, 701.
Anderson, S.F., Margon, B. & Grandi, S.A. (1983). *Ap. J.*, **273**, 697.

Bare, C.C., Clark, B.G., Kellermann, K.I., Cohen, M.H. & Jauncey, D.L. (1967). *Science*, **157**, 189.

Bartel, N., Herring, T.A., Ratner, M.I., Shapiro, I.I. & Corey, B.E. (1986). *Nature*, **319**, 733.

Barthel, P.D., Hooimeyer, J.R., Schilizzi, R.T. & Miley, G.K. (1987). In preparation.

Barthel, P.D., Miley, G.K., Schilizzi, R.T. & Preuss, E. (1985a). *Astron. Astrophys.*, **140**, 399.

Barthel, P.D., Schilizzi, R.T., Miley G.K., Jägers, W.J. & Strom, R.G. (1985b). *Astron. Astrophys.*, **148**, 243.

Biretta, J.A., Moore, R. & Cohen, M.H. (1986). *Ap. J.*, **308**, 93.

Blandford, R.D. & Königl, A. (1979). *Ap. J.*, **232**, 34.

van Breugel, W.J.M., Miley, G.K., Heckman, T.M., Butcher H.T. & Bridle, A.H. (1985). *Ap. J.* **290**, 496.

van Breugel, W.J.M., Nan, R., Venturi, T., Schilizzi, R.T., Fanti, C., Fanti, R. & Parma, P. (1987). In preparation.

Broten, N.W., Legg, T.H., Locke, J.L., Mcleish, C.W., Richards, R.S., Chisholm, R.M., Gush, H.P., Yen, J.L. & Galt, J.A. (1967). *Science*, **156**, 1593.

Browne, I.W.A., Clark, R.R., Moore, P.K., Muxlow, T.W.B., Wilkinson, P.N., Cohen, M.H. & Porcas, R.W. (1982). *Nature*, **299**, 788.

Burke, B.F. (1984). *Proceedings of the IAU Symposium 110 on VLBI and Compact Radio Sources* (eds R. Fanti, K.I. Kellermann & G. Setti), p. 397.

Carr, T.D., May, J., Olson, C.N. & Walls, G.F. (1965). *IEEE NEREM Record*, **7**, 222.

Cohen, M.H., Moffet, A.T., Romney, J.D., Schilizzi, R.T., Shaffer, D.B., Kellermann, K.I., Purcell, G.H., Grove, G., Swenson, G.W., Yen, J.L., Pauliny-Toth, I.I.K., Preuss, E., Witzel, A. & Graham, D. (1975). *Ap. J.*, **201**, 249.

Cohen, M.H., Unwin, S.C., Pearson, T.J., Seielstad, G.A., Simon, R.S., Linfield, R.P. & Walker, R.C. (1983). *Ap. J.*, **269**, L1.

Fanti, C., Fanti, R., Schilizzi, R.T., Spencer, R.E. & van Breugel, W.J.M. (1986). *Astron. Astrophys.*, **170**, 10.

Kellermann, K.I. & Thompson, A.R. (1985). *Science*, **229**, 123.

Levy, G., Linfield, R.P., Ulvestad, J.S., Edwards, C.D., Jordan, J.F., Di Nardo, S.J., Christiansen, C.S., Preston, R.A., Skjerve, L.J., Stavert, L.R., Burke, B.F., Whitney, A.R., Capallo, R.J., Rogers, A.E.E., Blaney, K.B., Maher, M.J., Ottenhoff, C.H., Jauncey, D.L., Peters, W.L., Nishimura, T., Hayashi, T., Takano, T., Yamada, T., Hirabayashi, H., Morimoto, M., Inoue, M., Shiomi, T., Kawaguchi, N. & Kunimori, H. (1986). *Science*, **234**, 187.

Miley, G.K. (1983). In *Astrophysical Jets* (eds A. Ferrari & A.G. Pacholczyk), Reidel, Dordrecht, p. 99.

Moran, J.M., Crowther, P.P., Burke, B.F., Barrett, A.H., Rogers, A.E.E., Ball, J.A., Carter, J.C. & Bare, C.C. (1967). *Science*, **157**, 676.

Norman, C.A. & Miley, G.K. (1984). *Astron. Astrophys.*, **141**, 85.

Porcas, R.W. (1987). *Proceedings of the Workshop on Superluminal Sources*, Pasadena, Oct. 1986 (eds J.A. Zensus & T.J. Pearson), p. 12.

Readhead, A.C.S., Hough, D.H., Ewing, M.S., Walker, R.C. & Romney, J.D. (1983). *Astrophys. J.*, **265**, 107.

Readhead, A.C.S., Masson, C.R., Moffet, A.T., Pearson, T.J., Seielstad, G.A., Woody, D.P., Backer, D.C., Plambeck, R.L., Welch, W.J., Wright, M.C.H., Rogers, A.E.E., Webber J.C., Shapiro, I.I., Moran, J.M., Goldsmith, P.F., Predmore, C.R., Baath, L.B. & Rönnäng, B.O. (1983). *Nature*, **303**, 504.

Readhead, A.C.S. & Wilkinson, P.N. (1978). *Ap. J.*, **223**, 25.

Rees, M.J. (1967). *Mon. Not. R. Astr. Soc.*, **135**, 345.

Reid, M.J. (1984). *Proceedings of the Workshop on QUASAT*, Gross Enzersdorf, June 1984, ESA SP-213, (ed. W.R. Burke), p. 181.

Reid, M.J., Schnepps, M.H., Moran, J.M., Gwinn, C.R., Genzel, R., Downes, D. & Rönnäng, B.O. (1987). *IAU Symposium 115 on Star Formation*, (eds M. Peimbert, J. Jugaku), p. 554.

Reid, M.J., Moran, J.M., Gwinn, C.R. (1987). *Proceedings of the Workshop on Radio Astronomy in Space*, Green Bank, October 1986. (ed. K.W. Weiler), p. 145.

Sagdeev, R.Z., Linkin, V.M., Blamont, J.E. & Preston, R.A. (1986). *Science*, **231**, 1407.

Schilizzi, R.T. (1987*a*). *Proceedings of the Workshop on Radio Astronomy in Space*, Green Bank, October 1986, (ed. K.W. Weiler), p. 153.

Schilizzi, R.T. (1987*b*). *Proceedings of the Workshop on Radio Astronomy in Space*, Green Bank, October 1986, (ed. K.W. Weiler), p. 165.

Schilizzi, R.T., Booth, R.S., Wilkinson, P.N., Preuss, E., Burke, B.F., Preston, R.A., Jordan, J.F. & Roberts, D.H. (1984). *Proceedings of the IAU Symposium 110 on VLBI and Compact Radio Sources* (eds R. Fanti, K.I. Kellermann & G. Setti), p. 407.

Schilizzi, R.T. & de Bruyn, A.G. (1983). *Nature*, **303**, 26.

Schilizzi, R.T., Miley, G.K., van Ardenne, A., Baud, B., Baath, L.B., Rönnäng, B.O. & Pauliny-Toth, I.I.K. (1979). *Astron. Astrophys.*, **77**, 1.

Thompson, A.R., Moran, J.M. & Swenson, G.W. (1986). *Interferometry and Synthesis in Radio Astronomy*, Wiley and Sons.

Vermeulen, R.C., Schilizzi, R.T., Fejes, I, Icke, V. & Spencer, R.E. (1987). In preparation.

de Waard, G.J. (1986). PhD Thesis, University of Leiden.

Wardle, J.F.C., Roberts, D.H., Potash, R.I. & Rogers, A.E.E. (1986). *Ap. J.*, **304**, L1.

Whitney, A.R., Shapiro, I.I., Rogers, A.E.E., Robertson, D.S., Knight, C.A., Clark, T.A., Goldstein, R.M., Marandino, G.E. & Vandenberg, N.R. (1971). *Science*, **173**, 225.

Wilkinson, P.N. & de Bruyn, A.G. (1984). *Mon. Not. R. Astr. Soc.*, **211**, 593.

Wilkinson, P.N. & de Bruyn, A.G. (1987). In preparation.

5

Radio-linked interferometers

J.G. DAVIES*

Summary

The development of radio-linked interferometry is discussed from a historical point of view.

Around 1950 radio sources were known as radio stars, and this reflected the idea prevalent at that time that they were galactic objects, and as stars were going to be of stellar diameter. The problems of building a radio Michelson interferometer with a baseline capable of resolving this were clearly enormous at that date, particularly bearing in mind that 125 MHz was a daringly high frequency to work at. In order to overcome the impossibility of maintaining phase coherence over hundreds of kilometres, Hanbury Brown, who had recently arrived at Jodrell Bank, devised the intensity, or post-detector, interferometer; and Roger Jennison, with M. K. das Gupta, an Indian student, constructed it, and tested the principle by measuring the diameter of the Sun, in the summer of 1951. Larger broadside arrays were then built and the instrument improved so that in 1952 the diameters of the Cygnus and Cassiopeia sources were measured. The sizes turned out to be only slightly less than the previously published lower limits. At about the same time in Australia, Bernard Mills had constructed a radio linked interferometer and measured the diameter of the Crab nebula, and found also a diameter of a few minutes of arc. So it seemed that very large baselines were not going to be required for many sources. In addition, the intensity interferometer suffered from the severe disadvantage that it would only work when the signal to noise ratio on each receiver was of the order of unity or greater. Thus with aerials of reasonable size, the list of suitable

*Professor Davies died 16 September 1988.

51

targets was already nearly exhausted, and at the same time phase stable Michelson interferometers had been developed at Cambridge and in Australia that could also measure diameters of the order of minutes of arc.

The 218 ft transit telescope had been completed at Jodrell Bank in 1947, and we saw that it could be used to give much greater sensitivity than was available anywhere else by using it as one end of a Michelson interferometer. Since in this case the effective signal-to-noise ratio depends only on the geometric mean of the two telescopes, and does not require that each has S/N of order unity, the second instrument could be relatively small and hence portable to different sites at varying distances from Jodrell Bank. It is interesting to note that the Michelson interferometer, developed and first used in optical astronomy on only a limited number of objects, has achieved a far greater success and resolution in the field of radio astronomy, while the intensity interferometer was developed and used in a similarly limited way in radio astronomy and has achieved its greatest use in the optical field. The intensity interferometer was also important because of the light it shed on theoretical physics. A number of eminent physicists said in effect that it was impossible, because light was photons, and how could a photon be detected simultaneously at two points miles apart!

Henry Palmer came to Jodrell Bank in 1952 and immediately started work on a series of instruments and observations that would steadily over the next two decades increase the resolving power of the interferometers from a few thousand wavelengths to 2 million wavelengths. This was achieved by operating on sites scattered over the country, and ending with the use of one of the R.R.E. telescopes at Defford, near Malvern, which we have since taken over as an essential element in the MERLIN system. Before 1957 when the Mk1 telescope was completed, the transit telescope was in use, and so observations were restricted to a region near the zenith. Even after that date the remote telescope was not steerable, and was usually used in transit, although it could be tipped in elevation by hand to view any source. Towards the end of the period a 25 ft steerable paraboloid, made of wire netting mounted on electrical conduit was used, and so could provide a variety of hour angles, but it was not until 1965 and the collaboration with R.R.E. that continuous tracking through all hour angles became possible. At the same time, the wavelength used was steadily decreased as new techniques became available. The technical problems involved in these operations were formidable. Delay tracking and fringe speed removing became necessary with ever increasing precision. Electro-mechanical fringe speed machines and acoustic delay lines of quartz and mercury were developed. These were controlled eventually by paper tape produced on a computer. In those days computers were not fast enough to be capable of performing the task on-line, and digital techniques of delay and phase rotation were out of the question. While a number of the sources were soon resolved, showing that, as expected at the time, their diameters were of the order of minutes of arc, there were also a considerable number that stubbornly refused to be resolved, and hence the continuous moves to longer baselines. The work was stimulated by the discovery of the quasars, and the resulting fact that a study of these objects would be of considerable cosmological as well as astrophysical importance. Throughout this time one baseline at a time was all that was considered, no doubt for reason of the great logistic difficulties of operating more than one at once. Even when both the Defford telescope and the Mk3 telescope at Wardle, near Nantwich were available, this remained so, even in the minds of those concerned, as is shown by the fact that in the document setting out the scientific case for the proposed Mk5 telescope of 400 ft diameter to be built at Meifod in Wales; although much stress is placed on the number of sources available to the Mk1A–Mk5 interferometer, only a single baseline is mentioned. The data obtained were compared with models of single or double sources, and the results presented as statistical distributions of sizes and types.

It was only when in 1974 it became clear that the Mk5 could not be funded, that Henry Palmer proposed that we should develop a system with a number of telescopes suitably placed at various distances from Jodrell Bank, and that all baselines should be correlated simultaneously. The proposal was to purchase three 25 m telescopes from E-systems, the firm in the USA that was at that time producing a series of 28 such telescopes for the VLA in New Mexico. Because of doubts expressed in certain quarters about the phase stability of the system, the project was funded in two stages. In the first, one telescope was built at Knockin in Shropshire, at a distance of 67 km from Jodrell Bank, together with a correlator and computer system capable of operating with the full array. When this was shown to be operating satisfactorily, funds were made available for two more telescopes at Darnhall and Tabley, 18 and 11 km respectively from Jodrell Bank.

It is worth interpolating a word about closure phase. The principle that if three telescopes are used, and all three baselines are correlated, then information about the astronomical, as opposed to the instrumental and atmospheric, contributions to phase can be deduced for the slight penalty that information concerning the absolute position of the source is lost, was clearly set out in 1958 by Roger Jennison. It has been said that we at Jodrell

Bank then forgot all about it, until we were reminded of it by its use elsewhere. I think that the truth is that our emphasis on collecting information on ever increasing baselines on a statistically significant sample of sources led us to consider only single baseline instruments, where closure is of no meaning.

This Multi Telescope Radio Linked Interferometer was originally known as the MTRLI, but subsequently has become universally referred to as MERLIN (replacing the word 'telescope' by 'element' in the acronym). Here phase has come into its own as a vital part of the technique of analysis. In many respects the analysis of the data can be regarded as the tail that has come to wag the dog of aperture synthesis. When we started out on MERLIN, we certainly did not expect to produce maps of the quality that have now become commonplace. This improvement is due almost entirely to developments in analysis produced by many people, notably Tim Cornwell, a student who came to us as MERLIN got under way and has now transferred his talents to the VLA, and to Roger Noble and Peter Wilkinson. Techniques and indeed programmes are no longer developed at any one place, but internationally, and are exchanged between observatories in many countries.

MERLIN has always been unique in filling a gap between the smaller cable-linked interferometers such as those at Cambridge and the VLA, and the intercontinental baselines covered by VLBI, the tape recording technique, although recently increased emphasis is placed on extending the analysis to include data obtained on the VLA, MERLIN and VLBI. Tom Muxlow and Bill Junor, one of our present students, have recently combined MERLIN data with EVN (Jodrell Bank, Cambridge, Westerbork, Bonn and Onsala telescopes) to produce a much more detailed picture of the Virgo source. This has a dynamic range of 2000 to 1, and still contains imperfections which will be further reduced as the software is improved.

What of the future? MERLIN is currently being extended to use one of the telescopes at Cambridge, and we expect to make the first observations in a few months. The present correlator will not be adequate to cope with six new baselines, and a completely new one is under construction. This will use 1200 specially designed correlator chips and 37 68000 processors, each of which is as powerful as any of our on-line computers today. We hope* to be able to build a new 32 m telescope at Cambridge since greater sensitivity and 24-hour coverage is required and to extend the system again to include the

*Since this article was written, SERC has funded this 32 m telescope.

telescope at Chilbolton in Hampshire. These extensions increase the gaps in the u–v coverage of the system, and we are investigating a new way of filling these gaps and hence avoiding the dangers of the analysis converging on a false solution. This is the process of Multi-frequency synthesis, and John Conway, another student at Jodrell Bank, is developing the analysis techniques required. By observing at several frequencies spread over a frequency range of say 25%, not only can the u–v gaps be filled in, but in principle a map of spectral index can also be determined.

6

Millimetre interferometry

D. DOWNES

Summary

The young science of millimetre astronomical interferometry has come into its own only during the current decade. The main millimetre arrays for the next five years will be those of Berkeley, Cal Tech, Nobeyama and IRAM (Institut de Radio Astronomie Millimétrique). From an instrumental point of view, the chief problems to overcome are sensitivity, which limits the maximum useful baseline, and hence the angular resolution, and the limited field of view. There is active interest in extending the existing mm-arrays with more antennas, and in building new millimetre and sub-millimetre interferometers. Current cost-sensitivity arguments tend to favour dish diameters in the 10–15-m range. There are strong incentives to extend the coverage of the existing arrays to higher frequencies.

Introduction

The universe accessible to millimetre interferometers is primarily that of cold ($T > 100\,\mathrm{K}$) matter, regions of high extinction ($A_v \sim 10$–1000 magnitudes), giant molecular clouds, bi-polar flows from forming stars, mass loss from evolved stars, and the atmospheres of the planets. The great advantage of the millimetre region over the centimetre range is the large number of molecular spectral lines; more than 600 have been detected in space so far. In particular, millimetre interferometers operating on the CO lines at wavelengths of 2.6 and 1.3 mm can provide detailed maps of the molecular gas in other galaxies. In the continuum, the millimetre interferometers can observe compact HII regions, quasars and stellar winds.

Thermal emission from dust has now been detected, for the first time with radio interferometers, at a wavelength of 3 mm with the Berkeley and Cal Tech arrays (Masson *et al.*, 1985; Plambeck *et al.*, 1985).

Millimetre interferometers: history and current instruments

The first astronomical interferometer at a wavelength of less than 1 cm was the Bordeaux solar interferometer (Delannoy, Lacroix & Blum, 1973), which operated at 8.6 mm with 2 × 2.5-m antennas. The Cambridge 5-km telescope was used for observations of Cygnus-A, W3 (OH) and a few other sources at a wavelength of 9 mm during the winters of 1979–80 (Alexander, Brown & Scott, 1984; Scott, 1981).

The first interferometer for the shorter millimetre wavelengths was the University of California, Berkeley interferometer at Hat Creek (Welch *et al.*, 1979), which had begun operation at 1.3 cm in the early 1970s and produced its first results on the SiO maser line at a wavelength of 3.4 mm in 1979. At that time, the interferometer consisted of two 6-mm dishes (Fig. 6.1), and has been expanded since then to 3 × 6 mm. This interferometer is equipped with a three-level spectral line correlator which allows 256 complex spectral

Fig. 6.1. Two of the three 6-m antennas of the University of California, Berkeley, millimeter interferometer at Hat Creek. The antennas move on tracks 300 m EW and 200 m NS.

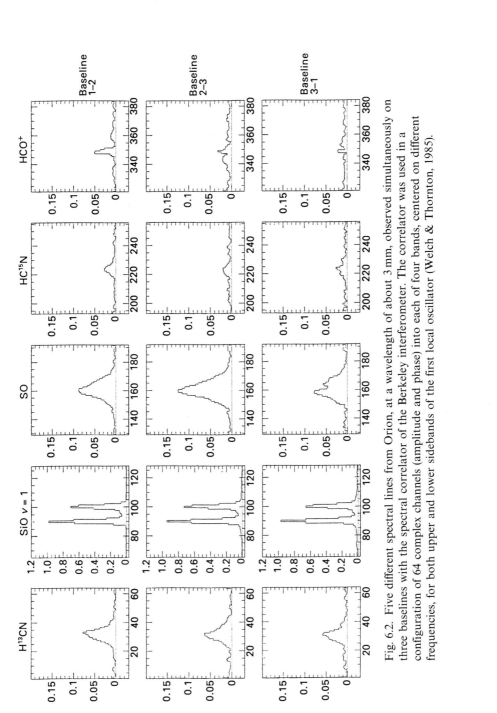

Fig. 6.2. Five different spectral lines from Orion, at a wavelength of about 3 mm, observed simultaneously on three baselines with the spectral correlator of the Berkeley interferometer. The correlator was used in a configuration of 64 complex channels (amplitude and phase) into each of four bands, centered on different frequencies, for both upper and lower sidebands of the first local oscillator (Welch & Thornton, 1985).

Fig. 6.3. Two of three 10.4-m antennas of the Cal Tech millimeter interferometer at Owens Valley. The antennas have surface errors $< 60\mu$m rms, and are usable to a wavelength of 1 mm.

channels per baseline pair, for each sideband. These may be split into four independent bands within the 500 MHz IF bandwidth, allowing observations such as are shown in Fig. 6.2, where five different spectral lines are observed on three baselines simultaneously (Welch & Thornton, 1985).

Fig. 6.4. Antennas of the 5 × 10-m array of the Nobeyama Radio Observatory. These antennas have surface errors < 70μm, and are usable to a wavelength of 1 mm.

The Cal Tech millimetre interferometer at Owens Valley, consisting of 3 × 10.4 mm antennas (Fig. 6.3) obtained its first fringes in 1980, in the continuum at a wavelength of 2.6 mm, and started CO line observations in 1982, as a two-element interferometer. Spectral line work with all three antennas has been going on since 1984 (Masson *et al.*, 1985). This array has pioneered interferometry of the CO line at 115 GHz in galactic and extragalactic objects.

The Nobeyama millimetre interferometer, with 5 × 10-m dishes (Fig. 6.4) started operating at 22 GHz in summer 1984, and conducted initial tests at 115 GHz in winter 1987–88. This interferometer has the longest baselines among the mm-arrays so far constructed, 560 m EW and 520 m at 33 deg from NS (Ishiguro *et al.*, 1985), and the largest number of baselines (10) which can be formed simultaneously.

The IRAM interferometer, under construction at Plateau de Bure, France, is to have 4 × 15-m dishes, located at 2550 m altitude and, as such, will have the largest antennas and the highest site of the mm interferometers built so far (Weliachew, 1985; Weliachew & Downes, 1985). The antenna technology differs from that used in radio astronomy dishes up to now, in that the panels and most of the reflector support structure are made of carbon fibre, rather

Fig. 6.5. The first of three 15-m antennas for the IRAM interferometer on Plateau de Bure, France. Initial surface measurements give a precision in the range 60—80μm rms. It is planned to commence with surface adjustments only when the second antenna is ready, in 1987, so that the surface can be checked with interferometric holography.

than metal, in order to ensure thermal stability (Delannoy, 1985). At present, the first two antennas have been completed (Fig. 6.5), and first rough measurements with tape measure and theodolite indicate an initial surface accuracy of 60–80 microns rms over the entire 15-m. The accuracy of individual panels (on a scale of approximately 1 m) ranges from 7 microns for the inner panels to 18 microns for the outermost ones. Tests were started in single-dish mode at a wavelength of 3.4 mm in winter 1986–87. The first fringes were obtained with 2 dishes late 1988.

Table 6.1 summarizes the properties of current millimetre arrays.

Differences relative to interferometers at cm wavelength

There are two ratios which are lower for millimetre than for centimetre arrays, namely,

$$\frac{\text{maximum baseline}}{\text{minimum baseline}} \quad \text{and} \quad \frac{\text{field of view}}{\text{size of typical sources}}.$$

Table 6.1. *Existing and potential mm-interferometers*

Array	n = no. of dishes	Instantaneous collecting area $= \pi n D^2/4$	Shortest wavelength (mm)	Relative speed for full synthesis
USA(Early NRAO idea) array: on one mount:	$20 \times 10\,\text{m} = 1570\,\text{m}^2$ $24 \times 3\,\text{m} = 170\,\text{m}^2$	$1740\,\text{m}^2$	1.0	63.3
Nobeyama 5 (in operation)	$5 \times 10\,\text{m}$	$393\,\text{m}^2$	1.0	3.3
Nobeyama 5 + 45 (in operation)	$1 \times 45\,\text{m} = 1590\,\text{m}^2$ $5 \times 10\,\text{m} = 393\,\text{m}^2$	$1983\,\text{m}^2$	2.0 to 3.0	5.0
Nobeyama (proposed)	$10 \times 10\,\text{m}$	$785\,\text{m}^2$	1.0	15.0
CSIRO (Australia telescope later phase)	$6 \times 10\,\text{m}$	$471\,\text{m}^2$	3.0	5.0
Cal Tech (in operation)	$3 \times 10.4\,\text{m}$	$255\,\text{m}^2$	1.3	1.0
Cal Tech (proposed)	$6 \times 10.4\,\text{m}$	$510\,\text{m}^2$	1.3	5.0
SAO-6 (proposed)	$6 \times 6\,\text{m}$	$170\,\text{m}^2$	0.35	5.0
Berkeley-3 (in operation)	$3 \times 6\,\text{m}$	$85\,\text{m}^2$	1.0	1.0
Berkeley-6 (funded)	$6 \times 6\,\text{m}$	$170\,\text{m}^2$	1.0	5.0
Plateau de Bure-3 (in construction)	$3 \times 15\,\text{m}$	$530\,\text{m}^2$	1.0	1.0
Plateau de Bure-4 (proposed)	$4 \times 15\,\text{m}$	$707\,\text{m}^2$	1.0	2.0
Plateau de Bure-6 (proposed)	$6 \times 15\,\text{m}$	$1060\,\text{m}^2$	1.0	5.0

In the first of these ratios, the minimum baseline is usually slightly larger than the size of the individual dishes, while the maximum usable baseline for typical sources studied may be limited by the sensitivity. For given integration times, bandwidths and noise limits, the maximum practical baseline, B_{max}, scales as $D(N/T_{sys})^{0.5}$, where D is the dish diameter, T_{sys} is the system temperature, and $N = n(n - 1)/2$ is the number of interferometer pairs formed with n dishes.

In comparison with most synthesis telescopes in the centimetre range, the millimetre arrays until now have had smaller and fewer antennas, higher system temperatures, more problems with the atmosphere, and weaker sources for both calibration and observing. For the most part, the thermal line sources which are observed in the millimetre range have brightness temperatures of only several degrees, and are weak in comparison with the non-thermal sources and nearly optically thick HII regions studied with interferometers in the centimetre range.

In practice, for the millimetre arrays until now, the minimum detectable brightness temperature in the synthesized beam reaches a few K at a baseline of about 100 m, and for fixed integration time, scales as the square of the baseline. However, the thermal spectral-line sources being studied have brightness temperatures of this order, so that it has been difficult so far to obtain useful signal-to-noise ratios on these sources at baselines much greater than 100 m. Indeed, most of the publications from the Berkeley and Cal Tech interferometers so far report maximum baselines of the order of 40–70 m, although these will probably be extended to 100 m in the near future.

The synthesized beamwidths corresponding to the typical baselines used so far at 3 mm are typically 6–11 arcsec, but these values are uncomfortably close to the sizes of structures for which information is missing from the maps, because of the minimum baseline. The graph in Fig. 6.6 shows the visibility amplitude for circular gaussian sources of full width to half-power from 2 arcsec to 2 arcmin. At a wavelength of 2.6 mm, a minimum spacing of 25 m, chosen to avoid mutual shadowing of antennas of, say, 15 m diameter, means that 9 arcsec structures will have their visibilities reduced by a factor of two, and structures > 20 arcsec will be missing completely from the interferometer maps.

In principle, the missing zero-spacing information can be obtained from single-dish maps. For the current mm interferometers, useful combinations have been and will be made with maps from the Onsala 20-m, the IRAM 30-m and the Nobeyama 45-m telescopes.

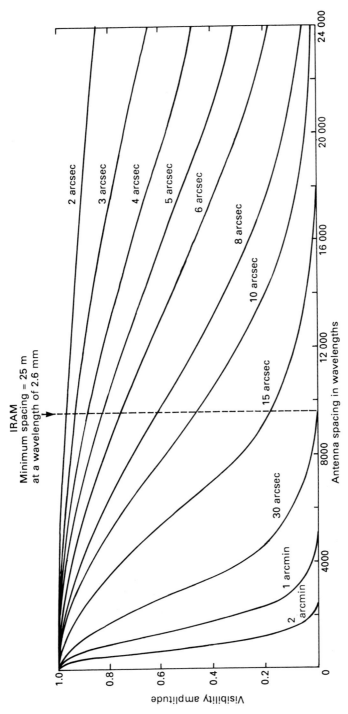

Fig. 6.6. Visibility amplitude of circular gaussian sources of diameter 2 arcsec to 2 arcmin. For example, for an array operating at 2.6 mm, with a minimum spacing of 25 m (9600 wavelengths), gaussian-like structures of half-power width 9 arcsec will have their fringe visibilities reduced by a factor of two. Structures > 20 arcsec will be completely missing from the interferometer maps.

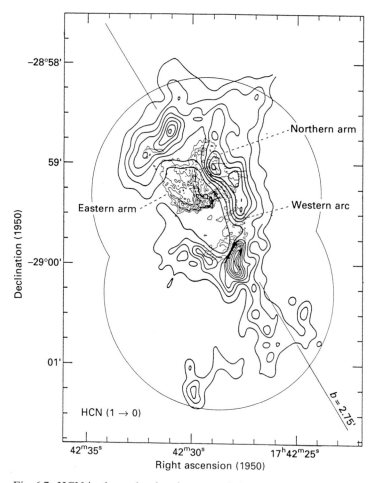

Fig. 6.7. HCN in the molecular ring around the center of the Galaxy, as mapped with the Berkeley mm interferometer (Güsten *et al.*, 1987). The region shown is larger than the primary beam of the 6-m antennas of the array, so two fields have been combined to make this map.

The second ratio mentioned above, for which the situation is more acute for millimetre arrays than for centimetre arrays, is the field of view, relative to the size of typical molecular-line sources. One would like to be able to map molecular clouds and nearby galaxies over an extent of at least several arcmin, but the primary beamwidth of current millimetre arrays is only about one arcmin at a wavelength of 3 mm, and the situation will become more difficult as some of these arrays start operating at the CO (2–1) line at 1.3 mm.

One way to overcome this problem will be to observe a mosaic of adjacent fields, and to combine the results into a single map in the data processing. Fig. 6.7 shows a successful application of this technique in the map of the 3-mm emission of the HCN molecule in the ring structure around the centre of the galaxy, as observed with the Berkeley mm interferometer (Güsten *et al.*, 1987). Another possibility might be to combine data taken with different interferometers. For example, the Hat Creek array might be used to do a full synthesis of a large field at shorter baselines, while the IRAM array could concentrate on the more compact sources in that field at longer baselines.

Plans for the future

There are proposals to expand all of the existing mm-arrays (Table 6.1). In the southern hemisphere, it is planned in a later phase of the Australia telescope to use the inner 10-m of the dishes in the 6×22-m compact array at Culgoora for observations at 3 mm. At the National Radio Astronomy Observatory, studies have begun for a large millimetre array (MMA). Early versions of this concept foresaw two instruments, to overcome the problems posed by the two ratios mentioned above: to have sufficient sensitivity to operate usefully at long baselines and thus to have high resolution, an instantaneous collecting area $> 1000 \, \text{m}^2$ is needed. This might be achieved with a large number (> 21) of reasonably large dishes ($> 10 \, \text{m}$ diameter). To obtain a large field of view for extended sources not requiring high resolution, the early MMA discussions foresaw an array of small (about 3-m) dishes, all mounted on a 25-m support structure.

A crucial factor in future millimetre arrays of large collecting area will be the cost. Fig. 6.8 shows a family of curves of cost vs. antenna diameter, based on recent experience. The curves are for given values of sensitivity (in the sense of signal-to-noise ratio), normalized to unity for a 6×6-m array. Because mm arrays will be sensitivity-limited for many applications, the observing speed of the array should perhaps be regarded as inverse of the time required to reach a specified noise level, rather than the number of baselines, as is usually done for arrays at centimetre wavelengths. Hence the relative observing speeds for the different arrays represented in Fig. 6.8 are given as the square of the relative sensitivities.

The cost curves in Fig. 6.8 increase rapidly for smaller dish sizes, where a larger number of dishes is required to achieve a given sensitivity, and hence the investment is dominated by the cost of the electronics rather than the antennas. The curves have broad flat regions for the larger dishes. From the cost-sensitivity analysis, the optimum dish diameter for high-sensitivity,

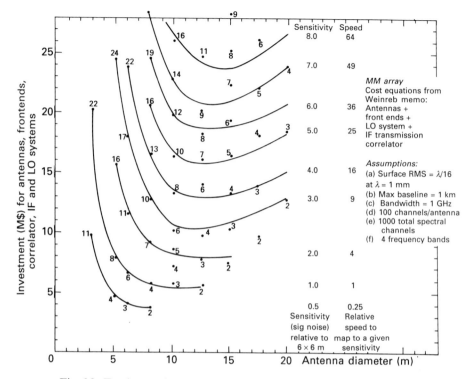

Fig. 6.8. Total cost of a millimetre interferometer vs. antenna size, for given sensitivity, normalized to a 6 × 6-m array. (From Downes, 1984, based on cost estimates for the electronics by Weinreb, 1983). Labels next to points indicate number of dishes of a given diameter. Integration times, aperture efficiencies, bandwidths, system temperatures and maximum baselines are assumed to be the same in all cases.

high-speed millimetre arrays appears to be in the 10–15-m range. As millimetre technology evolves in the future, the costs will change, but the general shape of the curves in Fig. 6.8 will probably remain the same.

In addition to the current proposals to extend the existing millimetre arrays, or to construct new ones, there is also the proposal of the Smithsonian Astrophysical Observatory to extend interferometry into the sub-millimetre region (Moran, 1985). Part of the motivation to do so is as follows: for molecular transitions, the Einstein A-coefficient varies as v^3; hence, for optically thin lines, the line brightness temperature varies as v^2, so the flux density in the line varies as v^4, and the integral of the line flux over frequency varies as v^5. For optically thick lines, the line flux will vary as v^2, and the frequency-integrated flux as v^3. For continuum emission from dust, the flux density will vary as v^3 or v^4, depending on the grain emissivity.

For millimetre interferometers, it may be of interest to vary the baselines as a function of frequency so as to obtain maps of a source in numerous spectral lines, but all at the same angular resolution. If one were to keep the angular resolution constant in this way, and also to keep constant resolution in velocity, then the brightness temperature sensitivity varies as $v^{2.5}$. Hence even if receivers are noisier, there are important advantages in observing at higher frequencies, and this will be an incentive to extend the coverage of millimetre interferometers.

Given the active interest in expanding the existing mm-arrays, in constructing new and more powerful ones, and the rewards to be obtained by moving to higher frequencies, the young field of millimetre astronomical interferometry obviously has a very promising future.

References

Alexander, P., Brown, M.T. & Scott, P.F. (1984). *Monthly Notices, Roy. Astron. Soc.*, **209**, 851.

Delannoy, J. (1985). In the ESO-IRAM-ONSALA Workshop on (Sub) Millimeter Astronomy, eds P. Shaver & K. Kjär, ESO, Garching, p 25.

Delannoy, J., Lacroix, J. & Blum, E.J. (1973). *Proc. IEEE*, **61**, 1282.

Downes, D. (1984). *MMA Memorandum No. 22*, National Radio Astronomy Observatory, Socorro.

Güsten, R., Genzel, R., Wright, M.C.H., Jaffe, D.T., Stutzki, J. & Harris, A.I. (1987). *Astrophys. J.*, in press.

Ishiguro, M., Morita, K.I., Kasuga, T., Kansawa, T., Iwashita, H., Chikada, Y., Inatani, J., Suzuki, H., Handa, K. & Takahashi, T. (1985). *In Proceedings of the International Symposium on Millimeter and Submillimeter Wave Radio Astronomy*, URSI, Granada, Spain, p. 65.

Masson, C.R., Berge, G.L., Claussen, M.J., Heiligman, G.M., Leighton, R.B., Lo, K.Y., Moffet, A.T., Phillips, T.G., Sargent, A.I., Scott, S.L., Woody, D.P. & Young, A. (1985). *In Proceedings of the International Symposium on Millimeter and Submillimeter Wave Radio Astronomy*, URSI, Granada, Spain, p. 65.

Masson, C.R., Claussen, M.J., Lo, K.Y., Moffet, A.T., Phillips, T.G., Sargent, A.I., Scott, S.L. & Scoville, N.Z. (1985). *Astrophys. J.*, **295**, L47.

Moran, J.M. (1985). *In Proceedings of the International Symposium on Millimeter and Submillimeter Wave Radio Astronomy*, URSI, Granada, Spain, p. 89.

Plambeck, R.L., Vogel, S.N., Wright, M.C.H., Bieging, J.H. & Welch, W.J. (1985). In *International Symposium on Millimeter and Submillimeter Wave Radio Astronomy*, URSI, Granada, Spain, p. 235.

Scott, P.F. (1981). *Monthly Notices, Roy. Astron. Soc.*, **194**, 25.

Weinreb, S. (1983). *MMA Memorandum No. 6*, National Radio Astronomy Observatory, Charlottesville.

Welch, W.J. & Thornton, D.D. (1985). In *Proceedings of the International*

Symposium on Millimeter and Submillimeter Wave Radio Astronomy, URSI, Granada, Spain, p. 53.

Welch, W.J., Dreher, J.H., Hoffman, W., Thornton, D.D. & Wright, M.C.H. (1979). *Astron. Astrophys.*, **59**, 379.

Weliachew, L.N. (1985). In *Proceedings of the International Symposium on Millimeter and Submillimeter Wave Radio Astronomy*, URSI, Granada, Spain, p. 85.

Weliachew, L.N. & Downes, D. (1985). In the ESO-IRAM-ONSALA Workshop on (Sub) Millimeter Astronomy, eds. P. Shaver & K. Kjär, ESO, Garching, P. 51.

7

Radio image processing

A.C.S. READHEAD

Summary

Recent developments in radio image processing are discussed and it is shown that these enable us to construct reliable images from observations with non-phase-stable non-uniform arrays.

1 Introduction

Developments in radio astronomical image processing over the last dozen years have revolutionized our whole approach to image construction in astronomy and opened the way to direct imaging of active galactic nuclei. The novel aspects of these new techniques consist primarily of methods for overcoming the well-known 'phase problem', familiar in many fields of physics, and the amplitude calibration problem; and of methods of compensating for limited sampling in the Fourier-transform domain. The application of these new image construction techniques is not limited to radio astronomy, but can be extended to any form of Michelson interferometry and to speckle interferometry.

The main driver for improving radio astronomical images has been the astrophysics, and some impressive recent astrophysical results are discussed in the article by Schilizzi in these proceedings. For the sake of brevity I shall concentrate here on image processing *per se.*

2 Limitations of radio image construction

Consider a radio interferometer consisting of two antennas labelled *m* and *n*. The signals from the two antennas are combined and give rise to the familiar interference fringes. The relationship between the observed correlated flux densities and phases, which we characterize by the complex variable $V_{mn}(t)$, and the true complex visibilities $\Gamma_{mn}(t)$ can be written:

$$V_{mn}(t) = G_m(t) G_n^*(t) K_{mn}(t) \Gamma_{mn}(t) + \alpha_{mn}(t) + \varepsilon_{mn}(t), \tag{1}$$

where $G_m(t)$ is the time-varying complex gain of antenna *m*; $K_{mn}(t)$ and $\alpha_{mn}(t)$ represent baseline dependent complex gains and offsets, respectively, which may be introduced, for example, by mismatches in the bandpasses or in the correlator; and $\varepsilon_{mn}(t)$ represents the thermal noise.

In general, the $K_{mn}(t)$ and $\alpha_{mn}(t)$ terms can be reduced to an acceptable level by careful instrumental design, and in aperture synthesis over short baselines the $G_m(t)$ can be determined accurately by calibration on celestial sources, so the major remaining source of error is the thermal noise $\varepsilon_{mn}(t)$. A major effort in phase-stable interferometry goes into the direct determination of the antenna gains, $G_m(t)$. Once this has been done, the images can be made by direct Fourier inversion, which is a *linear* combination of the measured Fourier components (Ryle & Hewish, 1960). The exploitation of this technique, first at Cambridge and then at Westerbork, made possible the detailed study of the physics of powerful extragalactic radio sources.

Ideally, we would always use phase-stable, well-calibrated systems. However, in practice there are many cases of great astrophysical interest for which the $G_m(t)$ cannot be determined directly. So the main challenge becomes the determination of these gain terms by indirect methods. The development of such methods has been the major achievement of radio imaging over the last dozen years. In addition to solving the problem of image formation in VLBI, these methods have greatly improved the image quality of phase-stable interferometers such as the VLA. Progress in this area depends on the use of two important pieces of information which are not used explicitly in classical aperture synthesis – namely, the *positivity constraint on the sky brightness distribution,* and the *finite extent of the source.*

In practice, interferometer measurements are made at a set of only *M* points (u_j, v_j) in the u–v plane, and thus the sky intensity and the visibility function must be approximated by the sum:

$$I(x, y) = \sum_{j=1}^{M} \Gamma(u_j, v_j) \, W_j \exp\left[2\pi i(u_j x + v_j y)\right], \tag{2}$$

where W_j is the 'weight' associated with the *j*th point. This gives an estimate, $I(x, y)$, of the sky brightness usually called the 'dirty map' which is a convolution of the true sky brightness, $O(x, y)$, with a point-spread function, or 'dirty beam' $P(x, y)$:

$$I(x, y) = O(x, y) \otimes P(x, y), \tag{3}$$

where $\otimes$ indicates a two-dimensional convolution. The dirty map is not the best possible representation of the sky. It contains artefacts such as the negative sidelobes around bright peaks, and if the sampling in the u–v plane is sparse some sidelobes will be large. This is due to the assumption in (2) that the visibility is zero in all unsampled regions of the u–v plane.

From the above discussion it should be clear that there are two major classes of problem that confront us in radio (and, for that matter, in optical) imaging:

(1) calibration of the observed quantities; and
(2) compensation for incomplete sampling in the u–v plane.

3 Indirect determination of the complex antenna gains

The calibration of the observed quantities is often extremely difficult or impossible, and this raises two questions.

(1) Do we have to eliminate both the amplitude and phase components of the baseline dependent errors, $G_{mn}(t)$, to make non-trivial images?
(2) Are accurate amplitude and phase calibrations always essential for image construction – i.e. do we have to be able to measure $G_m(t)$ *directly* to make non-trivial images?

The second question is particularly important in VLBI and in optical interferometry since the phase component of $G_m(t)$ cannot be determined at all by calibration.

We now know that the answer to (2) is *'no–there are many astrophysically important cases for which we can determine the complex gains indirectly and hence make good images'*. We are just beginning to explore (1), and I shall return to this later. For the present let us consider the indirect determination of the antenna based gains.

The first step towards the indirect determination of the phase of $G_m(t)$ was taken by Jennison in 1953 when he was a research student working under Hanbury Brown at Jodrell Bank. Jennison used three telescopes to eliminate random phase variations due to atmospheric and instrumental effects

(Jennison, 1953; 1958). Smith (1952) used three antennas to eliminate the unknown gain variation of one of the antennas; and in 1960 Twiss, Carter & Little used a four-telescope array to eliminate all antenna gain errors. These three pieces of work showed that it is possible to extract some information about the structure of the object even if $G_m(t)$ cannot be determined directly. However, these ideas lay fallow for many years, largely because the development of phase-stable interferometers was so successful that there seemed to be no need for these concepts. It turns out that this perception was incorrect.

The problems arose again in the development of VLBI. In VBLI the phase components of the gains, $G_m(t)$, can be calibrated directly only for a few objects for which a nearby bright point source can be used as a phase reference. In most cases the phases cannot be calibrated directly and thus it is not possible to make images by direct Fourier inversion. The solution to this problem is to use the *closure phase* (Rogers *et al.*, 1974; Wilkinson *et al.*, 1977; Readhead & Wilkinson, 1978). Similarly, the difficulty of calibrating visibility amplitudes accurately led to the use of the *closure amplitudes* (Readhead *et* al., 1980). The same result can be obtained by adjusting the complex antenna gains without explicit use of the closure phase or amplitude (Schwab, 1980; Cornwell & Wilkinson, 1981), and for many purposes this approach is preferable to the explicit use of the closure quantities. These methods of indirect determination of $G_m(t)$ have recently been reviewed by Pearson & Readhead (1984).

The closure phase is simply the sum of visibility phases around a closed loop of baselines. Since the smallest non-trivial case involves three telescopes, the simplest form of the closure phase, ψ_{lmn}, can be derived by applying equation (1) to three baselines between antennas l, m and n. We assume for the present that $K_{lm} = K_{mn} = K_{nl} = 1$, and $\alpha_{lm} = \alpha_{mn} = \alpha_{nl} = 0$, i.e. there are no baseline dependent errors other than thermal noise. Then we have:

$$\arg[V_{mn}(t)] = \arg[\Gamma_{mn}(t)] + \theta_m(t) - \theta_n(t) + \eta_{mn}(t), \tag{4}$$

where $\theta_m(t) = \arg[G_m(t)]$ is the phase of the complex gain, and $\eta_{mn}(t)$ is the phase error due to the thermal noise on baseline mn. Similar expressions apply for the other two baselines. Thus the closure phase defined by:

$$\begin{aligned}\Psi_{lmn}(t) &= \arg[V_{lm}(t)V_{mn}(t)V_{nl}(t)] \\ &= \arg[\Gamma_{lm}(t)\Gamma_{mn}(t)\Gamma_{nl}(t)] + \eta_{lm}(t) + \eta_{mn}(t) + \eta_{nl}(t), \end{aligned} \tag{5}$$

contains only terms involving the true visibility phase and the thermal noise, since the phase terms, arg $[\mathbf{G}_m(t)]$, all cancel. In practice the terms $\mathbf{K}_{lm}(t)$, etc., are not identically unity, nor are the $\alpha_{lm}(t)$ terms zero, so that there are some non-closing errors. However these are generally much smaller than the antenna based errors which are eliminated. Similarly, the closure amplitude, $A_{klmn}(t)$, is defined by:

$$A_{klmn}(t) = \frac{|\mathbf{V}_{kl}(t)|\,|\mathbf{V}_{mn}(t)|}{|\mathbf{V}_{km}(t)|\,\mathbf{V}_{ln}(t)|} = \frac{|\mathbf{\Gamma}_{kl}(t)|\,|\mathbf{\Gamma}_{mn}(t)|}{|\mathbf{\Gamma}_{km}(t)|\,|\mathbf{\Gamma}_{ln}(t)|}$$
$$+ \text{thermal noise terms},\qquad(6)$$

since the gain terms, $|\mathbf{G}_m(t)|$, all cancel.

Application of these concepts has led to a partial answer to question (2) and we now know that we can make images which are effectively thermal-noise limited from well-calibrated amplitudes and closure phases – i.e. we do not have to have accurately calibrated phases. In fact, even with fairly poorly calibrated amplitudes we can reach this limit by using the closure amplitudes as well as the closure phases. Most VLBI maps are made with fairly poorly calibrated amplitudes, and totally uncalibrated phases. A good example is the map of the nucleus of NGC 6251 by Jones *et al.* (1986), which is shown in Fig. 7.1. The noise level in regions of blank sky in this map is within a factor two of the thermal noise limit.

Even in the case where we have no calibration at all, fairly high dynamic range images can be made from the closure phases and closure amplitudes alone (Readhead *et al.*, 1980). Baldwin & Warner (1976) showed that it is possible to construct images from the visibility amplitudes alone in cases where the image consists of a number of isolated point sources, and we have recently carried out some tests at Caltech that show that it is possible to construct fairly complex images from phases alone.

A similar development to that in VLBI now taking place in optical speckle interferometry (Lohman, Weigelt & Wirnitzer, 1983; Weigelt & Wirnitzer, 1983) is discussed by Weigelt in these proceedings; and Baldwin discusses in these proceedings a procedure more directly parallel to that in radio astronomy which makes use of non-redundant aperture masks (see also Baldwin *et al.*, 1986). In speckle interferometry the observed image is again given by equation (3), and in this case the point spread function $P(x, y)$ varies rapidly with time due to atmospheric effects. The method of extracting phase information, which Weigelt and his collaborators have developed, involves the use of bi-spectra, which are triple products of complex visibilities on closed triangles of baselines (cf. equation (5)). Clearly, the phases of bi-spectra are simply closure phases: In cases where there are no

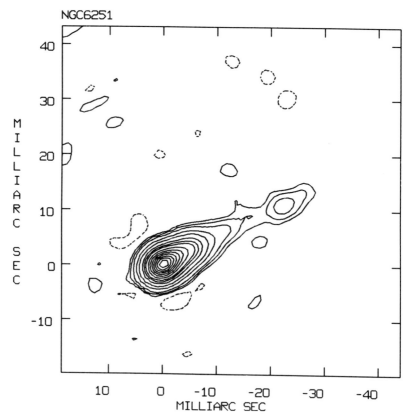

Fig. 7.1. The nucleus of NGC 6251 mapped with a VLBI network of 11 telescopes at a frequency of 1.7 GHz. This image was constructed from visibility amplitudes and closure phases and has a dynamic range (see text) of over 500. The noise in areas of blank sky is within a factor two of the thermal noise limit. The contour levels are -0.25, 0.25, 0.5, 1, 2, 5, 10, 15, 25, 35, 50, 70, 90% of the peak of 279 mJy/beam (from Jones *et al.*, 1985).

redundant spacings the contributions to the observed bi-spectra phases due to the $P(x, y)$ terms cancel exactly and, therefore, the bi-spectrum phase is simply equal to the closure phase of the object; whereas if there are redundant spacings a noise term is added to the object closure phase (Cornwell, 1986; Readhead *et al.*, 1988).

The problem of constructing an image in speckle interferometry therefore becomes similar to the problem of constructing an image from closure phases and amplitudes in radio astronomy. The optical amplitudes can be derived most directly from the modulus of the observed visibilities (Labeyrie, 1970), and the closure phases from the bi-spectra; and, since, in the case of full aperture coverage, all spacings are available up to the diameter of the mirror, it is possible to derive the visibility phases explicitly from the closure phases, once one phase has been assumed (Rogstad, 1968). A model can then be generated from the phase solution subject to the positivity constraint, and hence a new estimate for the initially assumed phase can be obtained. The procedure can then be iterated until a stable solution is found. This is the technique used by Lohman *et al.* (1983). It is very similar to the technique used at Westerbork, which likewise provides a lot of redundant spacings.

4 Image construction

We now consider the second major class of problems which confront us. In radio astronomy, particularly in VLBI, the coverage of the u–v plane is often sparse and/or non-uniform and this leads to large sidelobes and hence to very 'dirty' images. In order to construct the images it is necessary to make use of *non-linear* combinations of the measured quantities and thus to estimate or restore the unmeasured Fourier components. Much progress has been made in this area over the last dozen years (see, e.g. Pearson & Readhead, 1984; Narayan & Nityananda, 1986). Non-linear methods do a better job than linear methods because they do not set the visibility amplitudes to zero in unsampled regions of the u–v plane. However, the restoration is generally not unique, but relies on some other information, such as the positive brightness constraint and finite extent of the source, to select an acceptable solution from a continuum of possible solutions.

The most successful of these image construction techniques are the 'CLEAN' (Hogbom, 1974) and the 'MAXIMUM ENTROPY' (MEM) methods (Wernecke & D'Addario, 1977; Gull & Daniell, 1978).

4.1 Image construction by CLEAN

A good example of the power of CLEAN is shown in Fig. 7.2–a 22 GHz map of 3C84. In Fig. 7.2(*a*) is shown the dirty beam, and in Fig. 7.2(*b*) the dirty map. The CLEAN map is shown in Fig. 7.2(*c*). Clearly, the CLEAN procedure has led to a greatly improved image, but how close is this image to that which would be obtained with complete u–v coverage? In other words how well have we deconvolved the dirty map? The answer to this question depends not only on the u–v coverage but also on the signal-to-noise ratio and on the complexity of the source.

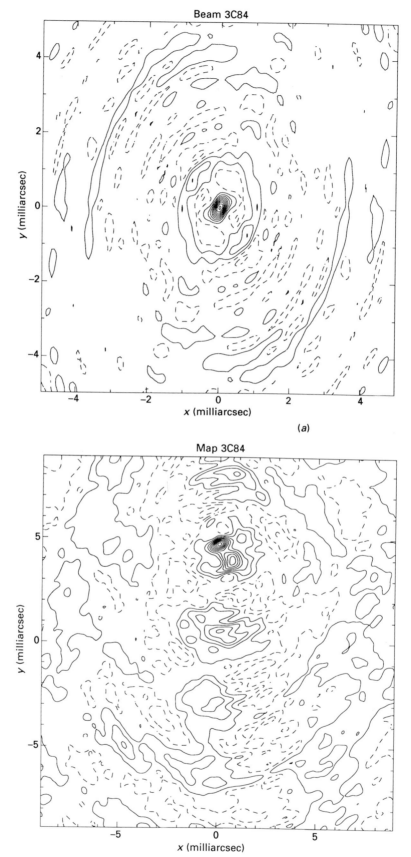

Fig. 7.2.

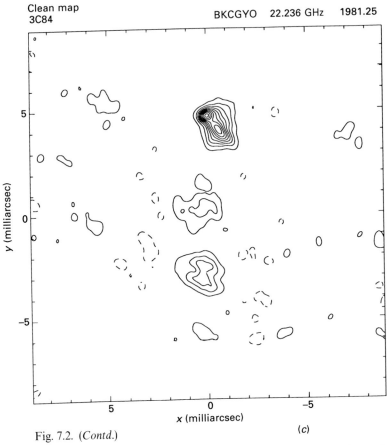

Fig. 7.2. (*Contd.*) (*c*)

Fig. 7.2. An example of CLEAN taken from 22 GHz VLBI observations of NGC1275 on a 6 station network: (*a*) the dirty beam; (*b*) the dirty map; (*c*) the CLEAN map. Note that the high-level sidelobes are eliminated in the CLEAN map. The contour levels are the same in all three cases, viz. $-15, -5, 5, 15, 25, 35, 45, 55, 65, 75, 85, 95\%$ of the peak brightness temperature of 5.7×10^{10} K. The beamsize is 0.5×0.35 milliarcsec^2 (from Readhead *et al.* 1983*a*).

In a series of simulations performed to test the imaging capability of an orbiting VLBI telescope ('QUASAT'), operating in conjunction with 10 ground stations, we measured the noise level in different regions of a CLEANed map as a function of the signal-to-noise-ratio (Readhead *et al.*, 1984). The results are shown in Fig. 7.3. We see that the noise level outside the emission regions is set entirely by the thermal noise – indicating that CLEAN has managed to subtract distant sidelobes of the dirty beam extremely well. However, in low brightness extended regions the noise is 3–4 times greater, and in the

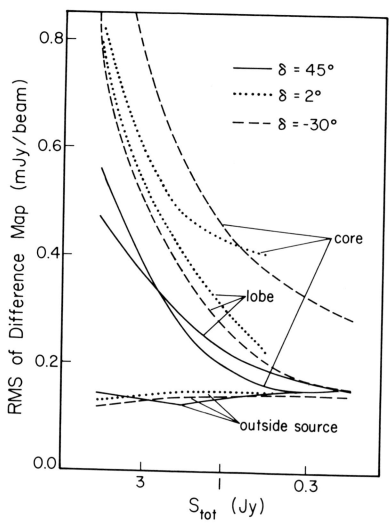

Fig. 7.3. Simulated noise levels on CLEAN maps made of a complex object with a space antenna operating in conjunction with 10 ground stations. Shown here are the rms differences between the simulated map and the original model, convolved with the same beam. The levels were calculated for objects with total flux densities of 5.6 Jy, 1.7 Jy, 0.56 Jy, and 0.17 Jy. The noise level on the maps in areas of blank sky is comparable to the thermal noise level, but in regions of extended emission or adjacent to bright peaks the sidelobes have not been completely suppressed, and discrepancies significantly greater than the thermal noise level are seen. These tests were performed for objects at three different declinations to illustrate the effect of variations in the uv coverage (from Readhead *et al.*, 1984).

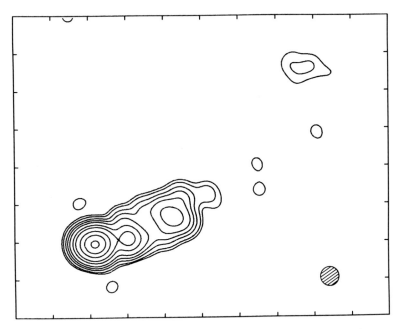

Fig. 7.4. 5-GHz map of 3C345. This example of the CLEAN restoration procedure has a dynamic range (see text) of about 2000. The contour levels are 0.1, 0.2, 0.5, 1, 2, 5, 10, 25, 50, 90% of the peak flux density. The tick marks are spaced 2.5 milliarcsec apart (from Biretta, 1985).

vicinity of bright, compact regions it is higher still. This figure illustrates the danger of using the term 'dynamic range' without defining it precisely. For the purpose of this discussion I define the dynamic range to be the ratio of the brightest feature on the map to the rms fluctuations on the map in a region of blank sky.

A further example of CLEAN is the 5 GHz map of 3C345 made by Biretta (1985) shown in Fig. 7.4. The noise in the regions of the map away from the bright regions is within a factor two of the thermal noise limit, and the dynamic range is about 1000. This map enables us to put stringent limits on the brightness level of any counter-jet, and hence provides useful constraints for the relativistic beaming theory.

Even in cases where we have a phase-stable and well-calibrated array of radio telescopes, the self-calibration and CLEAN or MEM procedures help to eliminate small systematic errors which would otherwise significantly reduce the dynamic range of the images. These techniques have therefore been applied at the VLA and at Westerbork and Cambridge.

One of the most stringent tests of CLEAN and self-calibration is the 5 GHz

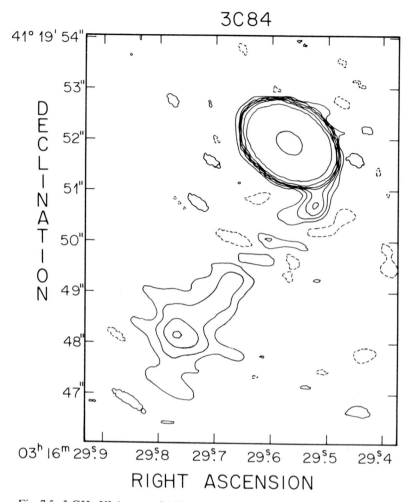

Fig. 7.5. 5-GHz VLA map of 3C84. The dynamic range of 1.8×10^5 in this map was achieved by careful calibration, self-calibration, and CLEAN (see text). The noise level in areas of blank sky is about a factor five above the thermal noise level. Small, non-closing baseline dependent errors are the limiting factor in this map. The large-scale structure visible here can only be detected by means of such very high dynamic range maps. This structure appears to be an extension of the extremely bright, one-sided structure seen in the VLBI maps (see Fig. 7.2(*c*)). The contour levels are -1, 1, 2, 3, 4, 5, 10, 50, 26350 mJy/beam, and the peak flux density is 52773 mJy/beam (Perley, private communication).

VLA map of 3C84 made by Perley (private communication) which is shown in Fig. 7.5. This was made using the VLA in the spectral line mode. Great care was taken to adjust the delays to within about 0.2 nanoseconds. The peak brightness on the map is 52.8 Jy/beam and the rms noise on the map in blank sky is 0.29 mJy/beam, giving a dynamic range of 1.8×10^5. This is about one-fifth of the dynamic range that would be obtained if the map were thermal noise limited, so it is clear that some non-closing errors remain. Without this very high dynamic range it would not be possible to see the one-sided extension to the south-east in this object. Note that the inner extension is in a similar position angle to the pc sized nuclear jet shown in Fig. 7.2(c) The southernmost component of the inner jet is moving away from the bright core at the north at a speed of $0.4c$, and presumably feeding the large scale jet seen in the VLA map.

4.2 Image construction by 'MEM'

The MEM method of image construction has now been used successfully in both conventional aperture synthesis and in VLBI. An advantage of MEM is that it naturally provides some degree of superresolution (Narayan & Nityananda, 1986) in cases of high SNR. It is thus often very useful in discriminating between features that are separated by less than the conventional beamwidth.

A superb example of the use of MEM and self-calibration is shown in Fig. 7.6–the 5 GHz VLA map of Cygnus A made by Perley and Dreher (private communication). There is so much information in this image that it is not possible to display it with a simple linear or logarithmic grey-scale plot. In order to display contrasting features at widely differing brightness levels Charles Lawrence worked together with the Image Processing Laboratory at JPL to do some sophisticated filtering of the image and produce the version shown in Fig. 7.6. The remarkable filaments and the jet, and possible counter-jet would be invisible without the use of self-calibration and CLEAN.

5 Global fringe fitting

In VLBI, owing to uncertainties in the propagation delay, the baseline parameters and the oscillator phase, the delay and fringe rate cannot be predicted accurately. Thus when the signals are cross-correlated a range of different delays and fringe rates are used. Traditionally, the fringe fitting is then done one baseline at a time. However, there is, of course, closure between the fringe rates and delays around closed loops of baselines, and thus closure can be

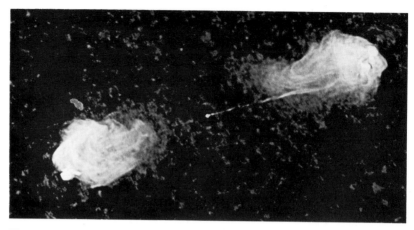

Fig. 7.6. MEM image of Cygnus A. This superb VLA image has a dynamic range of over 1000 and covers about 10^5 square beams. The high quality of the image reveals, for the first time, the channel along which matter travels to the western lobe, and numerous filaments in the outer lobes, as well as the detailed structure around the hot spots (Perley & Dreher, private communication).

used to narrow down the delay-fringe rate window on weak baselines. In fact the best approach is to do a global fit to the fringes on all baselines simultaneously (Schwab & Cotton, 1983) with weighting based on the signal-to-noise-ratio. For an unresolved source the closure delays and fringe rates are zero, but for a resolved source these have to be calculated.

The usual procedure is to start with a point source model and iterate two or three times. This is expensive in computer time but it does greatly improve the maps in cases where there are a number of weak baselines. The image shown in Fig. 7.1 was made by global fringe fitting.

6 Future imaging prospects in radio and optical astronomy

I now wish to discuss briefly the prospects for imaging in astronomy in the radio and optical frequency ranges. There are four important developments in the immediate future in radio astronomy:

(1) The completion of the VLBA;
(2) An orbiting radio telescope for VLBI (QUASAT);
(3) The extension of the MERLIN Array; and
(4) The exploitation of millimetre VLBI.

The VLBA will make possible, for the first time, the study of samples of objects at a number of frequencies and in different polarizations. This is vital for progress in this field. The orbiting telescope will complement both the VLBA and millimetre VLBI, and, by means of high dynamic range maps at 22 GHz, enable us to interpret our 90 GHz observations. The MERLIN extension will significantly increase the resolution and performance of this instrument and provide a vital link in baseline length between the present array and the European VLBI Network.

In order to make a complete physical interpretation of continuum observations it is essential to have observations over a wide frequency range with comparable resolution, and this can only be achieved with the combination of instruments listed above.

There have now been 5 successful 3 mm VLBI observing runs, and the results are very encouraging (Readhead *et al.*, 1983*b*; Rogers *et al.*, 1984). Coherence times of up to 200 seconds at 90 GHz have been observed, and in the last observing run we detected transcontinental fringes on all seven objects observed. The number of telescopes which can participate in these observations is small at present because most millimetre telescopes are not equipped with hydrogen maser frequency standards and Mark III recording terminals. However, within the next few years we expect the number of participating stations world-wide to increase to ten. With this array we will be able to make images at 90 GHz and 230 GHz with a resolution of 65 microarcsec and 25 microarcsec, respectively. Thus we will have a resolution of about 10^{16} cm in nearby active nuclei such as 3C84.

We turn now, finally, to the first question posed in Section 7.3. Is it possible to make non-trivial images in those cases in which even the closure amplitudes are corrupted? We have been looking into this problem recently at Caltech because we are interested in applying these methods both to millimetre wavelength VLBI and to optical interferometry. In both of these applications the amplitude calibration poses substantial difficulties. Not only are the amplitudes themselves difficult to calibrate due to the magnitude of propagation effects and instrumental errors, but in some cases the closure amplitudes are also corrupted. In optical interferometry, for example, the effects of vibration and of atmospheric dispersion are more severe for long baselines and, therefore, the closure amplitudes can be biased. However, the closure phases can still be measured, and these are not biased even when the signal-to-noise-ratio is low. If we cannot trust the closure amplitudes then the relevant questions are:

(1) can images be constructed from the closure phase alone; and if so
(2) can the thermal noise limit be reached in these images?

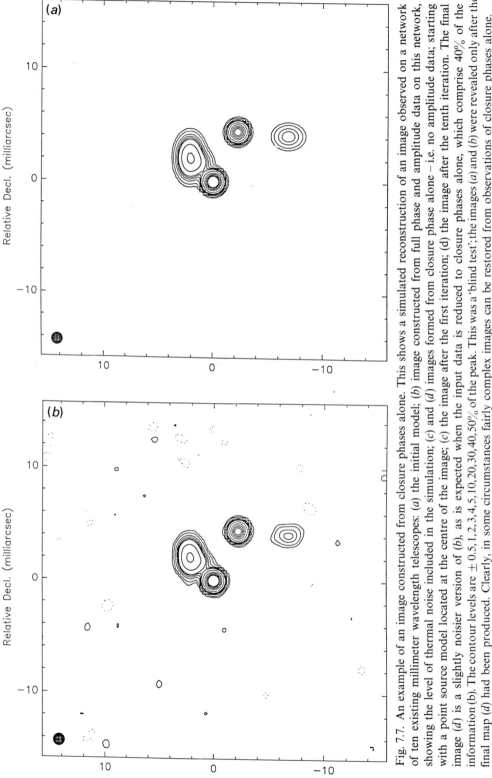

Fig. 7.7. An example of an image constructed from closure phases alone. This shows a simulated reconstruction of an image observed on a network of ten existing millimeter wavelength telescopes: (a) the initial model; (b) image constructed from full phase and amplitude data on this network, showing the level of thermal noise included in the simulation; (c) and (d) images formed from closure phase alone – i.e. no amplitude data; starting with a point source model located at the centre of the image; (c) the image after the first iteration; (d) the image after the tenth iteration. The final image (d) is a slightly noisier version of (b), as is expected when the input data is reduced to closure phases alone, which comprise 40% of the information (b). The contour levels are $\pm 0.5, 1, 2, 3, 4, 5, 10, 20, 30, 40, 50\%$ of the peak. This was a 'blind test'; the images (a) and (b) were revealed only after the final map (d) had been produced. Clearly, in some circumstances fairly complex images can be restored from observations of closure phases alone.

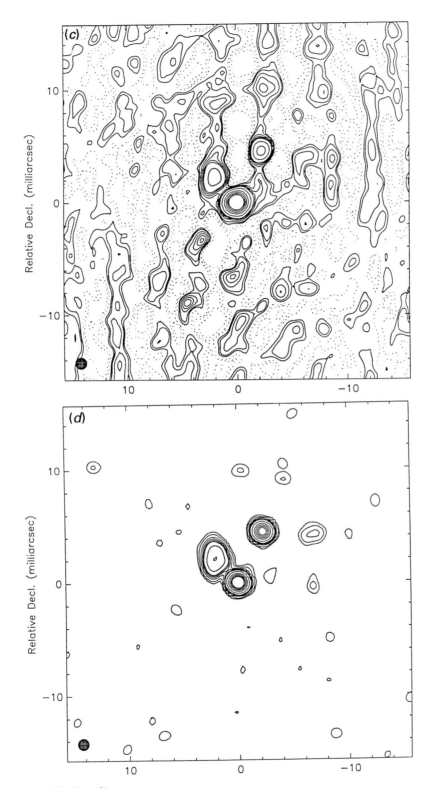

Fig. 7.7 (*Contd.*)

Over the last few months Tadashi Nakajima, Tim Pearson and I have been looking into this and we are quite encouraged by what we have found thus far. We have run a number of blind tests with very encouraging results.

Our procedure is as follows: we begin by assuming a point source model in the centre of the field. We set the amplitudes to unity and run an iteration of self-calibration using the observed closure phases. We then produce a dirty map and CLEAN this in the usual way. The output image is then used as the input model to the next stage. The amplitudes are again set equal to the model amplitudes and the observed closure phases are used in the self-calibration step of the second iteration. At each stage the amplitudes from the previous model are combined with the observed closure phases. Thus the closure amplitudes can change, but the closure phases are held constant.

The results of one of our tests are shown in Fig. 7.7. The original model is shown in Fig. 7.7(*a*). Figure 7(*b*) shows the image made with full amplitude and phase information, but with realistic thermal noise added. This is, therefore, a test of CLEAN with a ten station network. The thermal noise level is reached in areas of blank sky. In Fig. 7.7(*c*) we show the result of the first iteration, i.e. that with a point source input model. After ten iterations the model had converged to the image shown in Fig. 7(*d*). This is a very encouraging result. The noise level is about a factor two greater than in Fig. 7.7(*b*). This is about as good as could be expected since in this case there are only 36 independent observables (i.e. the 36 closure phases) compared with 90 independent observables (45 amplitudes and 45 phases) in Fig. 7.7(*b*). This also appears to be a useful way of producing a starting model for conventional VLBI imaging, and it could prove useful in the calibration of VLBI observations.

We suspect that images of extended objects with small brightness variations will be more difficult to recover, but, nevertheless, the results we have obtained thus far are very encouraging.

It appears that it should be possible to construct fairly complex images from both millimetre VLBI observations and optical interferometry. This raises the bright prospect of making VLBI observations of active nuclei with 25 microarcsec resolution and optical observations of active nuclei down to at least magnitude 15. In the case of the optical observations the continuum source might well be unresolved and therefore be usable as a phase reference which should enable us to map the broad emission line regions in these objects. The limiting resolution of the largest optical telescopes is 0.01 arcsec, and this resolution will be surpassed by long baseline optical interferometers, as discussed by Davis in these proceedings.

7 Conclusion

We have made significant progress in image construction with non-phase stable non-uniform arrays. At each step the difficulty or impossibility of proceeding by the usual methods has forced us to try alternative approaches, which have proven to be most useful even for phase stable uniform arrays. In order to make progress we have had to sacrifice some rigour and this has placed a greater burden on the observer who has to exercise considerable judgement in choosing the acceptable images from a large number of possible images. Nowadays, however, we are generally questioning features at the level of 0.1% to 1% of the peak brightness.

Finally, the exploitation of very high resolution interferometry at centimetre, millimetre, submillimetre, infrared and optical wavelengths is likely to be pivotal to many aspects of astrophysics over the next 20 years; and it is clear that the image processing methods developed in centimetre wavelength radio astronomy are directly applicable over this whole range of wavelengths.

References

Baldwin, J.E. & Warner, P.J. (1976). *MNRAS*, **175**, 345.
Baldwin, J.E., Haniff, C.A., Mackay, C.D. & Warner, P.J. (1986). *Nature*, **320**, 595.
Biretta, J.A. (1985). Ph.D. Thesis, California Institute of Technology.
Cornwell, T.J. & Wilkinson, P.N. (1981). *MNRAS*, **196**, 1067.
Cornwell, T.J. (1986). *Astron. Astrophys.* (in press).
Gull, S.F. & Daniell, G.J. (1978). *Nature*, **272**, 686.
Hogbom, J.A. (1974). *Astron. Astrophys. Suppl.*, **15**, 417.
Jennison, R.C. (1953). Ph.D. Thesis, University of Manchester.
Jennison, R.C. (1958). *MNRAS*, **118**, 276.
Jones, D.L., Unwin, S.C., Readhead, A.C.S., Sargent, W.L.W. & Seielstad, G.A. *et al.* (1986). *Ap. J.*, **305**, 684.
Labeyrie, A. (1970). *Astron. Astrophys.*, **6**, 85.
Lohman, A.W., Weigelt, G. & Wirnitzer, B. (1983). *Appl. Opt.*, **22**, 4028.
Narayan, R. & Nityananda, R. (1986). *ARAA*, **24**, 127.
Pearson, T.J. & Readhead, A.C.S. (1984). *ARAA*, **22**, 97.
Readhead, A.C.S. & Wilkinson, P.N. (1978). *Ap. J.*, **223**, 25.
Readhead, A.C.S., Walker, R.C., Pearson, T.J. & Cohen, M.H. (1980). *Nature*, **285**, 137.
Readhead, A.C.S., Hough, D.H., Ewing, M.S., Walker, R.C. & Romney, J.D. (1983*a*). *Ap. J.*, **265**, 107.
Readhead, A.C.S., Masson, C.R., Moffet, A.T., Pearson, T.J. & Seielstad, G.A., *et al.*, (1983*b*). *Nature*, **309**, 540.

Readhead, A.C.S, Preston, R.A., Meier, D.L., Linfield, R.P. & Lawrence, C.R. *et al.* (1984). Proceedings of Workshop on Quasat, Gr Enzersdorf, Austria, 18–22 June 1984 (ESA SP-213, September 1984) pp 101–110.

Readhead, A.C.S., Nakajima, T.S., Pearson, T.J., Neugebauer, G., Oke, J.B. and Sargent W.L.W. (1988). *Astron. J.*, **95**, 1278.

Rogers, A.E.E., Hinteregger, H.F., Whitney, A.R., Counselman, C.C. & Shapiro, I.I. *et al.* (1974). *Ap. J.*, **193**, 293.

Rogers, A.E.E, Moffet, A.T., Backer, D.C. & Moran, J.M. (1984). *Radio Science*, **19**, 1552.

Rogstad, D.H. (1968). *Appl. Opt.*, **7**, 585.

Ryle, M. & Hewish, A. (1960). *MNRAS*, **120**, 220.

Schwab, F.R. (1980). *Proc. Soc. Photo-Opt. Instrum. Eng.*, **231**, 18.

Schwab, F.R. & Cotton, W.D. (1983). *Astron. J.*, **88**, 688.

Smith, F.G. (1952). *Proc. Phys. Soc. London Sect.*, **B65**, 971.

Twiss, R.W., Carter, A.W.L. & Little, A.G. (1960). *Observatory*, **80**, 153.

Weigelt, G. & Wirnitzer, B. (1983). *Opt. Soc. America*, **8**, 389.

Wernecke, S.J. & D'Addraio, L.R. (1977). *IEEE Trans. Comput.*, **26**, 351.

Wilkinson, P.N., Readhead, A.C.S., Purcell, G.H. & Anderson, B. (1977). *Nature*, **269**, 764.

PART III
HIGH ANGULAR OPTICAL RESOLUTION

8

Quantum effects in high angular resolution astronomy

J.C. DAINTY

Summary

Theoretical analyses of the various techniques of high angular resolution astronomy, such as Michelson, intensity, speckle and phase-closure interferometry, are generally cast in a scalar wave treatment of light. In reality, one counts photo-electrons produced by a suitable detector. It is remarkable that the only effect of this quantum manifestation of light need be the lowering of the signal-to-noise ratio of a measurement. In fact, in some cases, measurements can be made at average rates of less than one photo-electron per sample time per (spatial) sample element.

1 Introduction

The use of optical interferometry to determine the spatial structure of astronomical objects was first proposed by Fizeau (1868). His double-slit interferometer was used by Stephen at the Marseille Observatory to observe 'all stars brighter than third magnitude', but he failed to resolve any stellar structure for the simple reason that the maximum slit separation was only 650 mm, corresponding to a resolution limit of approximately 0.16 arcsec in the visible range of the spectrum.

Michelson (1891) used an identical two-beam technique to resolve some of Jupiter's satellites. He later adapted it for baselines larger than a single telescope diameter and, with Pease (1921), obtained the first direct measurement of a stellar diameter (excepting the Sun), that of α-Orionis. Michelson's own description of his interferometer was based on simple scalar wave concepts of two beam interference, the fringe contrast, or

91

visibility, being reduced from the maximum value of unity if the star consists of more than one resolvable element. A modern description of this technique (and all other interferometric methods, see below) is cast in terms of coherence, or correlation, of scalar wave fields. Recent attempts to extend two beam amplitude interferometry to longer baselines are beginning to be successful (Blazit *et al.*, 1977; J. Davis, these proceedings) although the technical problems of maintaining path equalization are very great.

The 'intensity' interferometer of Hanbury Brown and Twiss (Hanbury Brown, 1984) is also invariably explained on a scalar wave model of light–in this case intensities, rather than amplitudes, of light interfere and in modern terms the relevant theory is that of fourth-order amplitude coherence functions (which are second-order intensity correlations). For the special, but commonly-occurring, case of a Gaussian light field, such as that emitted by a blackbody, the fourth-order correlation factors into sums of products of second-order correlations.

Speckle interferometry (Labeyrie, 1970; Dainty, 1984) and the optical phase-closure technique (Baldwin *et al.*, 1986) also are based on a scalar wave treatment of light. As with Michelson and intensity interferometry, the language of Fourier (wave) optics is used to explain how each technique works and of course the final detected signal is assumed to be the intensity of the scalar wave.

As all observers know, the most efficient detectors in the visible and near-infrared part of the spectrum produce discrete photo-electrons as their output, rather than a current proportional to intensity as in the radio wave region. Historically, this apparent dilemma was first discussed in connection with the intensity interferometer (Brown & Twiss, 1956; 1957) and gave a significant impetus to research on the quantum aspects of light (Glauber, 1963; Mandel & Wolf, 1965). It is a remarkable fact that, provided one processes the data correctly, the *only* effect of the quantum nature is to reduce the signal-to-noise ratio of the measurement.

2 Models of light

For a proper understanding of the nature of light and its interaction with matter, one obviously wants to understand the most detailed – and 'correct' – model that is currently available. This is based on quantum mechanics and assumes that the electromagnetic field is quantized (Glauber, 1963; Mandel & Wolf, 1965). This model is essential to explain a number of phenomena encountered in laser physics. Of special interest to astronomers might be the phenomenon of photon anti-bunching (Short & Mandel, 1983), which can only be explained on a quantized field model.

Because of the complexity of the quantized field model, it is natural to seek simpler models which are adequate to solve the problem at hand. For example, the lens designer almost always uses a ray picture of light, which is also the limit of the wave picture as the wavelength goes to zero. As mentioned above, all techniques of stellar interferometry rely on a scalar wave approximation to describe their operation (ignoring quantum effects). Optical scattering problems often require an electromagnetic wave description of light, interacting classically with electrons and nuclei in materials (Lorentz model).

If we are to introduce quantum effects into our understanding of the various techniques of stellar interferometry, then the most sophisticated model we need is the so-called *semi-classical* model first described by Mandel (1959): the *propagation* of fields is described by classical wave theory (scalar or vector as necessary) and *detection* by a Poisson process in which the rate function is proportional to the classical intensity. That is, in the photocathode of a photomultiplier for example, the probability that a photoelectron will be emitted in a certain time interval is proportional to the integrated intensity over the same time interval.

Fig 8.1 illustrates this model for two cases, (a) uniform classical intensity and (b) the classical intensity associated with a thermal light source. Mandel (1959) showed that the probability of counting n photoelectrons in time T, $p(n:T)$, is related to the probability density of integrated classical intensity, $p(W)$ by

$$p(n:T) = \frac{1}{n!} \int_0^\infty p(W)e^{-W} W^n \, dW, \tag{1}$$

where W is related to intensity $I(t)$ through

$$W \propto \int_t^{t+T} I(t) \, dt. \tag{2}$$

For case (a) in Fig. 8.1, $p(W) = \delta(W - \bar{n})$, where $\bar{n}$ is the mean photoelectron count and therefore the photo-electron counting distribution reduces to the well-known Poisson distribution:

$$p(n:T) = \frac{e^{-\bar{n}} \bar{n}^n}{n!}, \tag{3}$$

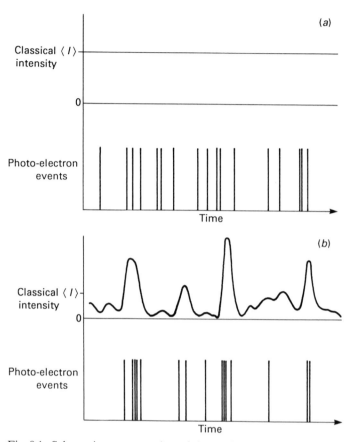

Fig. 8.1. Schematic representation of the semi-classical model: (*a*) constant classical intensity gives rise to photo-electrons with homogeneous Poisson statistics, (*b*) thermal light intensity results in a compound Poisson process with a Bose–Einstein counting distribution and apparent photo-electron 'bunching'.

whose variance $\sigma_n^2 = \bar{n}$. Note that this is the minimum possible variance according to the semi-classical model, as it corresponds to a classical intensity that shows no fluctuation at all. Sub-Poissonian light (i.e. $\sigma_n^2 < \bar{n}$) would require that the intensity had a negative variance, which is clearly unacceptable in the semiclassical picture.

On the other hand, thermal light (case (*b*) in Fig. 8.1) has an instantaneous intensity with a negative exponential probability density function

$$p(I) = \frac{1}{\bar{I}} e^{-I/\bar{I}} \tag{4}$$

and provided one samples in a time T short compared to the coherence time (this may be difficult in practice), the photo-electron counting distribution is Bose-Einstein

$$p(n:T) = \frac{\bar{n}^n}{(\bar{n}+1)^{n+1}}. \tag{5}$$

with variance $\sigma_n^2 = \bar{n} + \bar{n}^2$. The occurrence times of the photo-electrons now appear to be bunched: this is a direct result of the fact that their probability of emission is proportional to the classical intensity.

It should be stressed that the semi-classical model deals with the experimental observable, i.e. electrons emitted as a result of exposure to light. It makes no mention of 'photons', which are a result of quantizing the field. Despite the fact that the standard textbooks on quantum mechanics refer to the photoelectric effect as evidence for the quantized field (i.e. photons), virtually all aspects of it are explainable on the semi-classical model (Mandel, 1976)! It is invariably assumed, however, that photo-electrons are simply 'detected photons' and we shall use this terminology for the remainder of this paper.

3 Photon correlation

The technique of intensity and speckle interferometry rely on the ability to estimate correlations of instantaneous classical intensities at two or more points in space. In practice, the received signals are photo-electron counts averaged over space and time and the question immediately arises as to whether the quantized nature of the signal prevents an unbiased estimate of the intensity correlation being made.

Let $N_1(t)$ and $N_2(t)$ be the numbers of photons detected (i.e. photo-electrons emitted) by photodetectors 1 and 2 between time t and $t + \Delta t$. If $\Delta t \ll \tau_c$, where τ_c is the coherence time of the radiation, then it is a remarkable result that (Mandel, 1959)

$$\frac{\langle N_1(t) N_2(t+t_1) \rangle}{\langle N_1 \rangle \langle N_2 \rangle} = \frac{\langle I_1(t) I_2(t+t_1) \rangle}{\langle I_1 \rangle \langle I_2 \rangle} \tag{6}$$

where I_1 and I_2 are the classical instantaneous intensities.

Equation (6) states that the normalized correlation of photon counts is equal to that of the instantaneous intensity (provided the detectors are distinct). That is, photon correlation is an unbiased estimator of the intensity correlation. Signal processing in stellar interferometry is frequently carried out in the Fourier domain where the energy or power spectrum is calculated and a similar correspondence between energy spectra of photon counts and intensity also exists.

In case where the photodectors at points 1 and 2 are not distinct and where the time lag $t_1 = 0$, there is a constant 'photon noise' bias term which can be calculated from the data and subtracted. An example of this is presented in Section 8.4. In the case where triple correlations are involved, the bias term is more complicated (Lohmann *et al.*, 1983) but the vector-differencing algorithms used to calculate photon counting correlations can automatically allow for this bias by ignoring all self-differences of photon coordinates (Dainty & Northcott, 1986).

Thus the only effect of quantum fluctuations is to limit the accuracy with which a measurement can be made. Unfortunately, the signal-to-noise ratio (s/n) of each technique of steller interferometry involves different formulae and so it is difficult to make general statements as to the magnitude of errors introduced by quantum fluctuations. However, if one is able to sample the underlying intensity with negligible spatial and temporal integration (this is possible in most techniques with the exception of the Hanbury Brown intensity interferometer), then reliable estimates of the intensity correlation are possible with less than one detected photon per sample time per sample element and typically $10^4 - 10^5$ samples.

4 Example: s/n in speckle interferometry

In speckle interferometry, the average energy spectrum of the short exposure image, $\phi_1(u, v) = \langle |i(u, v)|^2 \rangle$, is estimated, where $i(u, v)$ is the Fourier transform of the image intensity $I(x, y)$. The normalized energy spectrum is related to that of the object by (Dainty, 1984)

$$\phi_1(u, v) = \phi_0(u, v) \cdot \langle |T(u, v)|^2 \rangle, \tag{7}$$

where $\langle |T(u, v)|^2 \rangle$, the speckle transfer function is given by

$$\langle |T(u, v)^2 \rangle = |\langle T(u, v) \rangle^2 + \frac{T_D(u, v)}{N_s} \tag{8}$$

and $T_D(u, v)$ is the diffraction-limited transfer function of the telescope and N_s is the number of speckles in the image.

To investigate the effect of quantum fluctuations, we model the jth image, $D_j(x, y)$, as an inhomogeneous Poisson point process with a rate proportional to the classical image intensity $I_j(x, y)$, i.e.

$$D_j(x, y) = \sum_{k=1}^{N} \delta(x - x_{jk})\delta(y - y_{jk})$$

where the delta functions represent detected photon events (x_{jk}, y_{jk}) is the location of the kth photon in the jth frame and N_j is the number of detected photons in the jth frame. The ensemble averaged squared modulus of the Fourier transform of this data is given by (Dainty & Greenaway, 1979)

$$\langle |d_j(u,v)|^2 \rangle = \bar{N}^2 \phi_i(u,v) + \bar{N}, \tag{9}$$

where $\bar{N}$ is the average number of detected photons per frame.

Using the estimator $Q = |d_j(u, v|^2 - N_j$, we therefore find that the expectation of Q is (Dainty & Greenaway, 1979)

$$\langle Q \rangle = \bar{N}^2 \phi_1(u, v)$$

and its variance σ_Q^2 is

$$\sigma_Q^2 = \bar{N}^2 + \bar{N}^2 \phi_1(2u, 2v) + 2\bar{N}^3 \phi_1(u, v) + \bar{N}^4 \phi_1^2(u, v) \tag{10}$$

For high frequencies, the second term in Eq. (10) can be ignored, yielding a s/n per frame of

$$s/n = \frac{\bar{N}\phi_1(u, v)}{1 + \bar{N}\phi_1(u, v)} \tag{11}$$

Finally, using Eqs (7) and (8) and defining $\bar{n}$ as the average number of detected photons per speckle, the s/n per frame becomes

$$s/n = \frac{\bar{n}T_D(u, v)\phi_0(u, v)}{1 + \bar{n}T_D(u, v)\phi_0(u, v)} \tag{12}$$

For very bright objects, $nT_D\phi_0 \gg 1$, the s/n approaches unity – one of the disadvantages of the speckle technique is that the s/n cannot exceed unity, regardless of the light level. For very faint objects

$$s/n \sim \bar{n}T_D(u, v)\phi_0(u, v), \tag{13}$$

that is, the s/n is directly proportional to the number of detected photons (and not its square root, as in conventional imaging).

An example of the variation of s/n per frame as a function of the average number of detected photons per frame $\bar{N}$ is shown in Fig. 8.2 for values of $D/r_0 = 10, 20$ and 40 (D is the telescope diameter and r_0 is Fried's seeing parameter) and $T_D \cdot \phi_0 = 0.05$. The horizontal broken line represents a s/n

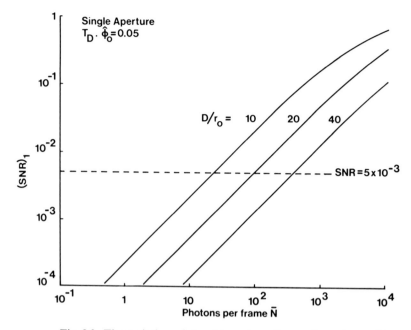

Fig. 8.2. The variation of signal-to-noise ratio per frame in speckle interferometry with the average number of detected photons per frame $\bar{N}$ for $D/r_0 = 10, 20$ and 40 and $T_D\phi_0 = 0.05$.

per frame of 5×10^{-3}: that is, 10^6 statistically independent frames (roughly one night of observation) would yield an s/n of 5 at each independent point in the power spectrum. For a large telescope ($D/r_0 \sim 40$), we require several hundred detected photons per frame and the corresponding magnitude turns out to be on the order of $m_v = 13$ (see Dainty, 1984, for details).

The above magnitude limit is for a good ($s/n \sim 5$ at $T_D \cdot \phi_0 = 0.05$) estimate of the whole object power spectrum. If one only wishes to estimate a single parameter, e.g. diameter or binary separation and position angle, then the limiting magnitude can be roughly 100 times fainter. Fig. 8.3 shows the result of a calculation by Walker (1981) for a 2000 sec observing period for the estimation of a diameter and its results are consistent with reported observations (Hege *et al.*, 1981; Arnold *et al.*, 1979) of binaries and multiple systems at the level of $m_v \sim 16$–17.

5 Conclusions

Quantum effects in high angular resolution astronomy can be understood completely in terms of the semi-classical picture of light

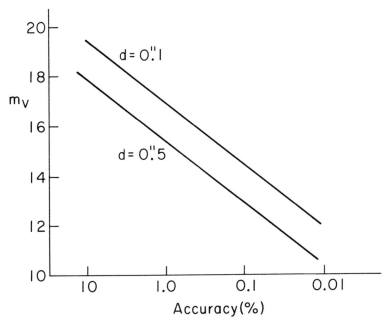

Fig. 8.3. Limiting magnitude m_v as a function of the desired fractional accuracy of an estimate of the diameter for a typical set of observing parameters on a 4-m class telescope and an observing period of 2000 sec.

developed by Mandel, in which one deals with photo-electrons rather than photons (quantized field). Quantum effects need not bias the results: they only influence the signal-to-noise ratio.

References

Arnold, S.J. *et al.* (1979). *Astrophys. J.*, **234**, L159.

Baldwin, J.E. *et al.* (1986). *Nature*, **320**, 595.

Blazit, A. *et al.* (1977). *Astrophys. J.*, **217**, L55.

Dainty, J.C. (1984). In *Laser Speckle and Related Phenomena*, 2nd edn, Springer-Verlag, Heidelberg.

Dainty, J.C. & Greenaway, A.H. (1979). High Angular Resolution Stellar Interferometry, *Proc. IAU Colloq.*, No. 50, University of Sydney.

Dainty, J.C. & Northcott, M. (1986). *Opt. Commun.*, **58**, 11.

Fizeau (1868). *Compt. Rend.*, **66**, 932.

Glauber, R.J. (1963). *Phys. Rev.*, **130**, 2529 and **131**, 2766.

Hanbury Brown, R. (1984). *The Intensity Interferometer*, Taylor and Francis, London.

Hanbury Brown, R. & Twiss, R.Q. (1956). *Nature*, **178**, 1447.

Hanbury Brown, R. & Twiss, R.Q. (1957). *Nature*, **179**, 1128.

Hege, E.K. *et al.* (1981). *Astrophys. J.* **248**, L1.

Labeyrie, A. (1970). *Astron. Astrophys.*, **6**, 85.

Lohmann, A.W. *et al.* (1983). *Appl. Opt.*, **22**, 4028.

Mandel, L. (1959). *Proc. Phys. Soc. (Lond.)*, **74**, 233.

Mandel, L. (1976). In *Progress in Optics*, Vol 13 (ed E. Wolf), North Holland, Amsterdam.

Mandel, L. & Wolf, E. (1965). *Rev. Mod. Phys.*, **37**, 231.

Michelson, A.A. (1891). *Nature*, **45**, 160.

Michelson, A.A. & Pease, F.G. (1921). *Astrophys. J.*, **53**, 249.

Short, R. & Mandel, L. (1983). *Phys. Rev. Lett.*, **51**, 384.

Walker, J.G. (1981). *Opt. Acta.*, **28**, 885.

9

The Sydney stellar interferometry programme

JOHN DAVIS

Summary

The Chatterton Astronomy Department of Sydney University was founded in 1961 to build and operate the Stellar Intensity Interferometer at Narrabri in northern New South Wales. After the programme with the Narrabri instrument was successfully completed in the mid-1970s the Department undertook the development of a successor. As a major step towards this goal a prototype of a modern form of Michelson's stellar interferometer has been built and tested. Based on experience gained in developing the prototype a very long baseline instrument has been designed. Construction of this new high angular resolution instrument will commence in 1987.

1 Introduction

The main theme of this meeting is intended to be forward-looking but the programme of stellar interferometry at the University of Sydney owes its origins to the intensity interferometer. It would therefore be a serious omission in view of Hanbury Brown's principal role if the historical background to the current stellar interferometry programme at Sydney was not outlined first, albeit briefly.

Stellar interferometry effectively came into being with the measurement of the angular size of Betelgeuse (Alpha Ori) by Michelson & Pease (1921), but it had a relatively short life at that time. The problems posed by atmospheric turbulence and by the necessity for extreme mechanical stability were overwhelming and could not be overcome with the technology available in the 1920s.

The field remained dormant until the discovery and development of intensity interferometry by Hanbury Brown and Twiss, and its application to stellar astronomy with their classic measurement of the angular size of Sirius (Alpha CMa) (Hanbury Brown & Twiss, 1956). It was this measurement that lead to the building of the Narrabri Stellar Intensity Interferometer and to the foundation of the Sydney programme.

2 The Narrabri Stellar Intensity Interferometer

The Narrabri Stellar Intensity Interferometer (Hanbury Brown, Davis & Allen, 1967) is shown in Fig. 9.1. The instrument had two 6.5 m diameter reflectors mounted on carriages running on a 188 m diameter circular railway track. In operation the line joining the two reflectors, the baseline of the interferometer, was maintained constant in length and

Fig. 9.1. An aerial view of the Narrabri Stellar Intensity Interferometer showing the two 6.5 m diameter reflectors on the 188 m diameter circular railway track and the central laboratory building.

Table 9.1. *Achievements with the Narrabri Stellar Intensity Interferometer*

1. Determination of the angular diameters of 32 stars
   – leading to the temperature scale for stars hotter than the Sun
2. Study of the double-lined spectroscopic binary Alpha Vir
   – determination of distance, primary mass, radius and luminosity, and secondary mass
3. Performance of a number of exploratory experiments
   – measurement of the angular size of emission envelope for Gamma 2 Vel
   – limb darkening for Alpha CMa
   – extended corona for Beta Ori
   – rapid rotation for Alpha Aql

perpendicular to the direction to the star throughout an observation. The maximum baseline of the instrument was set by the diameter of the railway track, namely 188 m.

Although the Narrabri instrument was limited in sensitivity to stars hotter than the Sun and brighter than $B = + 2.5$ the programme carried out with it produced results of considerable interest and importance. The temperature scale for stars hotter than the Sun is now firmly based on the 32 angular diameters measured at Narrabri (Code *et al.*, 1976), and the detailed study of the double-lined spectroscopic binary Spica (Alpha Vir) (Herbison-Evans *et al.*, 1971) demonstrated the power of interferometry for empirically determining fundamental stellar properties (mass, radius and luminosity). Table 9.1 summarizes the achievements of the observational programme with the Narrabri instrument, including some exploratory experiments. These latter experiments could not be pursued to the level of astrophysical significance because of the lack of sensitivity but they nevertheless demonstrated, in no uncertain manner, the potential of high angular resolution stellar interferometry.

3 A successor to the Narrabri Stellar Intensity Interferometer

At the completion of the Narrabri programme it was clear that there was an excellent case for building a more sensitive interferometer. The questions we asked ourselves were, firstly, what sensitivity, resolution and accuracy were required to make a significant contribution to astrophysics and, secondly, what type of interferometer should we build and how sensitive could it be made? In a sense these are chicken and egg questions! Dealing with the second question first, there are basically three types of optical interferometer to be considered:

> (i) An *intensity interferometer*–has the advantages of being essentially unaffected by the atmosphere and having large dimensional tolerances set by the radio frequency correlation bandwidth (100 MHz at Narrabri) which makes the engineering relatively easy. It has the significant disadvantage of being relatively insensitive and requires very large light collectors.
>
> (ii) A *small-aperture amplitude interferometer* – as pioneered by Twiss (Tango & Twiss, 1980). The 'small aperture' refers to the fact that the apertures are limited to the size of a single seeing cell and have diameters typically ~ 10 cm. This approach deals with the effects resulting from atmospheric turbulence in real time and, although technically more difficult than an intensity interferometer because of the extremely tight tolerances set by the wavelength of light and by the optical bandwidth used, it promises both sensitivity and accuracy.
>
> (iii) A *large aperture amplitude interferometer* – as pioneered by Labeyrie (1975). This approach promises greater sensitivity than either an intensity or a small-aperture amplitude interferometer but it is not expected to compete with them in accuracy. It has the same tight tolerances as the small-aperture amplitude interferometer.

Of these approaches only the last appears capable of measuring extragalactic objects. The first two techniques lack the necessary sensitivity but, on the other hand, both promise greater accuracy. We decided at an early stage in our programme that accuracy was vital if we were to make significant contributions to stellar astrophysics and this was our primary aim. We therefore carried out a study of a number of potential programmes to establish targets of sensitivity, resolution and accuracy.

Table 9.2 contains a list of some potential stellar programmes for a high angular resolution stellar interferometer (see Davis, 1979, 1983 and McAlister, 1979 for detailed discussions of some of these programmes). Table 9.3 lists the targets for sensitivity, resolution and accuracy that we set as a result of our study of these programmes.

Our initial choice, made in 1971, was to build a very large intensity interferometer because it was the only proven technique and the perfor-

Table 9.2. *Some potential programmes for a high resolution stellar interferometer*

1. Emergent fluxes and effective temperatures of stars
2. Radii and luminosities of stars of known distance
3. Binary stars – distances
                   – masses, radii and luminosities of primaries
4. Variable stars – Cepheid and Mira distances
                   – radii and luminosities
5. Emission line stars – size and shape of continuum and emission line sources
6. Rotating stars
7. Limb darkening
8. Interstellar extinction

Table 9.3. *Design targets for a high angular resolution stellar interferometer*

Parameter	Target
Sensitivity	V (limit) $> +7$
Baseline (max)	~ 300 m (except for early B and O stars)
	$(\theta_{min} \sim 2 \times 10^{-4}$ arcsec at 450 nm)
	> 600 m (for a sample of early O stars)
	$(\theta_{min} < 1 \times 10^{-4}$ arcsec at 450 nm)
Accuracy	$< \pm 2\%$

mance could be extrapolated from the Narrabri instrument with no unknown factors apart from potential improvements in detectors. However, by the time funding was granted to carry out a detailed design study there had been a number of developments which led us to reconsider. It became clear that a modern version of Michelson's stellar interferometer, in the form of a small-aperture amplitude interferometer, offered greater sensitivity at lower cost than the large-intensity interferometer that we had considered initially. It was obvious from the outset that the construction of such an instrument would be a major feat of optical engineering, involving active optics, laser interferometry, state of the art optical components, etc., and we decided that it would be prudent to develop a prototype interferometer as a first step. A prototype interferometer has therefore been built and its performance has been successfully demonstrated.

In parallel with the construction of the prototype, and based on experience gained in its development, a new very long baseline stellar interferometer has been designed and costed. Construction of this new instrument will commence in 1987.

4 The basic design

The basic features of the design of both the prototype and the new very long baseline interferometer follow directly from our desire, based on astrophysical considerations, to measure angular diameters, etc., to $\pm 2\%$ or better.

An amplitude interferometer is sensitive to both spatial and temporal variations in the phase of the interfering wavefronts. In the small-aperture approach that we have adopted, external phase errors resulting from turbulence in the Earth's atmosphere are mitigated by limiting the aperture diameters to the order of r_0, the transverse scale length for phase fluctuations (Fried & Mevers, 1974), and by an active optical servo to maintain beam alignment to within ~ 0.1 arcsec. Even with these precautions atmospheric turbulence will reduce the observed visibility (Tango & Twiss, 1980) so the atmospheric MTF must be continuously monitored and a correction applied. This is done via a specially developed 'seeing' monitor (O'Byrne, 1985).

Internal phase errors in the stellar interferometer are minimized by supporting the main optics on massive concrete foundations and by the use of optical components of the highest quality. The residual aberration in the instrument is corrected after the entire optical train has been assembled by means of specially figured optical plates. Moving components (in particular, the plane input mirrors and the path compensating retroreflectors) are carefully engineered to minimize phase jitter. Additional optics are incorporated in the design so that the internal alignment of the interferometer can be monitored.

5 The prototype interferometer

An aerial view of the prototype interferometer, which is located in the grounds of Australia's National Measurement Laboratory near Sydney, is shown in Fig. 9.2.

The prototype interferometer, which has been described in some detail by Davis & Tango (1985*a*), includes all the essential features of the very long

Fig. 9.2. The prototype modern Michelson stellar interferometer. The siderostat at the northern end of the 11.4 m north–south baseline can be seen on the left of the picture. The southern siderostat is hidden by the central laboratory building which houses the interferometer. On the right of the picture is a 0.35 m telescope which carries a real-time seeing monitor.

baseline instrument. In fact, the beam-combining optical system has been designed so that it can be transferred directly to the long baseline instrument. The prototype has 15 cm diameter siderostat mirrors at each end of a fixed 11.4 m north–south baseline to direct starlight into the instrument, an internal optical path compensation system, active wavefront-tilt correction, and the signals are rapidly sampled so that the effects of phase fluctuations are minimized.

Correlation (corresponding to the square of fringe visibility) was first observed with the prototype in October 1985 within ~ 10 minutes of starting a scan with the optical path compensator. Subsequently, the instrument has been used to determine the angular diameter of Sirius in a clear demonstration of the potential of the technique (Davis & Tango, 1986). This represents the first measurement of a main-sequence star by amplitude interferometry and it is also the first independent check of the measurements made with the Narrabri Stellar Intensity Interferometer. The new result is in excellent agreement with the Narrabri value (within 0.5% with a formal error of $\pm$ 1.4%), and took only ~ 2% of the observing time for the Narrabri result. This decrease in observing time, while impressive, does not truly represent the sensitivity of the instrument since only a very narrow optical bandwidth (0.3 nm) was used, the time included the measurement of complete delay curves for Sirius and a reference star, and, finally, neutral density filters were used to reduce the light flux from Sirius!

The prototype interferometer is currently engaged in an extensive observational programme which is aimed at fully understanding the influence of seeing on its performance and in optimizing its performance for the prevailing seeing conditions.

6 The very long baseline stellar interferometer (VLBSI)

The design of the VLBSI has been described by Davis & Tango (1985*b*). The principal differences between the new instrument and the prototype are that the VLBSI will have larger siderostat mirrors of diameter 20 cm, baselines ranging from 5 m to 640 m, and it will be located at a dark

Fig. 9.3. An artist's impression of the proposed new stellar interferometer looking towards the south–west. Part of the 640 m long north–south baseline array is shown in the foreground. The spacings and sizes are not to scale and the siderostat and vacuum pipe covers are not depicted. The laboratory complex which houses the optical path length compensator and beam-combining optics etc. is offset 50 m to the west of the baseline array.

site with a high incidence of clear skies and good seeing. An artist's impression of the new instrument is shown in Fig. 9.3.

The basic specification of the new interferometer includes 11 fixed siderostat stations, linked by a system of evacuated pipes, along a north–south line 640 m long. Beam switching optics will allow any siderostat station along the north arm to be combined with any station from the south arm to give a range of baselines increasing from 5 m to 640 m in steps of ∼ 40%. The shortest baseline will allow some overlap with measurements made using speckle interferometry on the largest optical telescopes, and the longest baseline has been chosen to enable a sample of the hottest stars to be measured. The clear aperture size of the instrument will be 14 cm when allowance is made for the projection of the 20 cm diameter siderostat mirrors.

The path compensation required for the longest baselines is formidable. It will be achieved by two moveable carriages travelling on 70 m long tracks with multiple folding of the beams to give a total range of ± 420 m. The positions of the carriages will be monitored by a laser interferometer system. A fringe locating and tracking system is under development and will be used to track the zero order fringe. This is essential for baselines longer than a few tens of metres (Davis, 1984).

Apart from the range of baselines, larger siderostats, and longer path compensator, the major instrument differs from the prototype in having provision for two sets of beam-combining optics on separate optical tables. Beam splitters will direct blue light to one and longer wavelengths to the other. The 'blue' beam-combining optics from the prototype will be installed on one of the tables for the first observing programmes with the new instrument. Future developments include beam-combining optics for other spectral ranges in the red and near infrared for installation on the second optical table to provide wide spectral cover. Only a single narrow spectral band (< 1 nm) is currently used in the prototype interferometer but a multi-spectral channel detection system will be used in the VLBSI. This will not only improve the signal-to-noise ratio for correlation measurements but will also allow simultaneous observations in the continuum and in spectral lines.

After an extensive survey, a site near Culgoora in northern New South Wales has been chosen for the new instrument. It will be built alongside the Australia telescope, which is currently under construction, at the centre of a flat, clear area approximately 1.5 km square. Geological, vibrational and engineering studies have shown that the site is close to ideal. Sky conditions are excellent.

Care has been exercised in the design of the new instrument and in the choice of site to ensure that potential future developments are not excluded. For example, the site will permit future extension of the north–south baseline to more than a kilometre and the addition of an east–west arm. The latter will allow east–west and diagonal baselines to be employed to improve the position angle coverage for the study of objects such as rotating stars and binary stars. The future use of the instrument for large-angle astrometric measurements, requiring the addition of a vacuum optical path compensator and additional laser metrology to monitor internal path lengths, will be possible. It will also be possible to bring three separate beams into the instrument from the north, south and east arms for phase-closure observations. However, these are for the future. Initially the instrument will have only the 640 m long north–south array of baselines and it will tackle the programmes listed in Table 9.2.

The cost of the VLBSI, based on detailed estimates, is A$2.39 million (April, 1985). The Australian Minister for Science has announced funding (October, 1986) through the Australian Research Grants Scheme to allow construction of the new instrument to start in 1987. Since much of the optics already exists in the prototype, and because of the experience gained in installing and aligning them, it is estimated that the construction time for the interferometer will be only $2\frac{1}{2}$ years.

The Narrabri Stellar Intensity Interferometer effectively opened up the field of high angular resolution stellar interferometry. The new very long baseline stellar interferometer will extend this work in a wide range of astrophysical programmes, many of which cannot be tackled in any other way, and it will be a worthy successor to the Narrabri instrument.

Acknowledgements

It is a particular pleasure in this context to acknowledge my good fortune in having worked closely with Emeritus Professor Hanbury Brown for the past 25 years in Australia. He initiated the Sydney programme of stellar interferometry, led it with great distinction from 1961 until 1981 and continues to support the programme with great enthusiasm. It is also a pleasure to acknowledge the contributions of Drs L.R. Allen, R.A. Minard, J.W. O'Byrne, R.R. Shobbrook, W.J. Tango and R.J. Thompson to the development of the prototype interferometer. The programme would not have been possible without the support of an initial feasibility grant by the

Australian Government, the Australian Research Grants Scheme, the University of Sydney Research Grants Committee, the Durack Trust Fund and the Science Foundation for Physics within the University of Sydney.

References

Code, A.D., Davis, J., Bless, R.C. & Hanbury Brown, R. (1976). *Astrophys. J.*, **203**, 417.

Davis, J. (1979). In *High Angular Resolution Stellar Interferometry* (*I.A.U. Colloquium No. 50*), eds J. Davis & W.J. Tango, p. 1–1. Chatterton Astronomy Department, University of Sydney.

Davis, J. (1983). In *Current Techniques in Double and Multiple Star Research* (*I.A.U. Colloquium No. 62*), eds R.S. Harrington & O.G. Franz, p. 191. Lowell Observatory Bulletin No. 167.

Davis, J. (1984). In *Indirect Imaging* (ed. J.A. Roberts), p. 125 (Cambridge University Press).

Davis, J. & Tango, W.J. (1985*a*). *Proc. astr. Soc. Australia*, **6**, 34.

Davis, J. & Tango, W.J. (1985*b*). *Proc. astr. Soc. Australia*, **6**, 38.

Davis, J. & Tango, W.J. (1986). *Nature*, **323**, 234.

Fried, D.L. & Mevers, G.E. (1974). *Appl. Opt.*, **13**, 2620 (correction in *Appl. Opt.*, **14**, 2567).

Hanbury Brown, R., Davis, J. & Allen, L.R. (1967). *Mon. Not. R. astr. Soc.*, **137**, 375.

Hanbury Brown, R. & Twiss, R.Q. (1956). *Nature*, **178**, 1046.

Herbison-Evans, D., Hanbury Brown, R., Davis, J. & Allen, L.R. (1971). *Mon. Not. R. astr. Soc.*, **151**, 161.

Labeyrie, A. (1975). *Astrophys. J.*, **196**, L71.

McAlister, H.A. (1979). In *High Angular Resolution Stellar Interferometry* (*I.A.U. Colloquium No. 50*), eds J. Davis & W.J. Tango, p. 3–1. Chatterton Astronomy Department, University of Sydney.

Michelson, A.A. & Pease, F.G. (1921). *Astrophys. J.*, **53**, 249.

O'Byrne, J.W. (1985). *Proc. astr. Soc. Australia*, **6**, 43.

Tango, W.J. & Twiss, R.Q. (1980). In *Progress in Optics*, ed. E. Wolf, **XVII**, 239.

10

Speckle masking, speckle spectroscopy and optical aperture synthesis[†]

G. WEIGELT

Summary

The atmosphere of the earth restricts the resolution of conventional astrophotography to about 1 arcsec. Much higher resolution can be obtained by using speckle methods. The speckle masking method yields images of general astronomical objects with diffraction-limited resolution, for example, 0.03 arcsec resolution for a 3.6 m telescope. True images are obtained since speckle masking reconstructs both the modulus and the phase of the object Fourier transform. Therefore, speckle masking is a solution of the phase problem in speckle interferometry. The limiting magnitude of speckle masking is about 20 mag. Speckle spectroscopy is a speckle method that yields objective prism spectra with diffraction-limited angular resolution. Finally, speckle masking can reconstruct high-resolution images from optical long-baseline interferograms. A 1-km telescope array on the earth would yield images with 10^{-4} arcsec resolution. With a 40-km array in space a fantastic resolution of 10^{-6} arcsec can be achieved at $\lambda \sim 200$ nm. We show speckle masking observations of NGC 3603 and Eta Carinae and computer simulations of optical aperture synthesis.

1 Introduction

In 1970 Antoine Labeyrie invented the speckle interferometry method (Labeyrie, 1970). This method yields the autocorrelation of astronomical objects with diffraction-limited resolution in spite of image

† Based on data collected at the European Southern Observatory, La Silla, Chile.

113

degradation by the atmosphere. The speckle masking method (Weigelt, 1977; Weigelt & Wirnitzer, 1983; Lohmann, Weigelt & Wirnitzer, 1983) has the additional advantage that it yields *true images* with diffraction-limited resolution. Speckle masking can be applied to general objects. A point source in the isoplanatic patch is not required. Speckle interferometry and speckle masking overcome both image degradation caused by the atmosphere and image degradation caused by telescope aberrations. The limiting magnitude of both techniques is about 20 mag. References of other speckle techniques are given in the review article by C. Dainty (1985).

In Section 10.2 we discuss the principle of speckle masking and speckle spectroscopy and we show speckle masking applications. In Section 10.3 optical aperture synthesis is discussed and illustrated by computer simulations.

2 Speckle masking and speckle spectroscopy

In speckle masking the same speckle raw data are evaluated as in speckle interferometry. The intensity distribution $I_n(\mathbf{x})$ of the nth recorded speckle interferogram can be described by

$$I_n(\mathbf{x}) = O(\mathbf{x}) * P_n(\mathbf{x}) \qquad n = 1, 2, \ldots, N \tag{1}$$

where $\mathbf{x}$ is a 2-dimensional vector, $O(\mathbf{x})$ is the 2-dimensional object intensity distribution, $P_n(\mathbf{x})$ is the instantaneous pointspread function of atmosphere/telescope, and $*$ denotes the convolution operator. Speckle masking processing consists of three steps:

Step 1. Calculation of the

triple correlation $I_n^{(3)}(\mathbf{x}, \mathbf{x}')$

or the

bispectrum $\tilde{I}_n^{(3)}(\mathbf{u}, \mathbf{v})$

of all speckle interferograms $I_n(\mathbf{x})$:

$$I_n^{(3)}(\mathbf{x}, \mathbf{x}') := \int I_n(\mathbf{x}'') I_n(\mathbf{x}'' + \mathbf{x}) I_n(\mathbf{x}'' + \mathbf{x}') d\mathbf{x}'' \tag{2}$$

$$\tilde{I}_n^{(3)}(\mathbf{u}, \mathbf{v}) := \int \int I_n^{(3)}(\mathbf{x}, \mathbf{x}') \exp[-2\pi i(\mathbf{u} \cdot \mathbf{x} + \mathbf{v} \cdot \mathbf{x}')] d\mathbf{x} \, d\mathbf{x}'$$

$$= \tilde{I}_n(\mathbf{u}) \tilde{I}_n(\mathbf{v}) \tilde{I}_n(-\mathbf{u} - \mathbf{v}). \tag{3}$$

The tilde $\sim$ denotes Fourier transformation and **u** and **v** are the coordinates in Fourier space. The bispectrum is the generalized Fourier transform of the triple correlation. Eqns (2) and (3) show that the triple correlation and the bispectrum are 4-dimensional functions of coordinates $(\mathbf{x}, \mathbf{x}')$ or $(\mathbf{u}, \mathbf{v})$, respectively, if the speckle interferograms $I_n(\mathbf{x})$ are 2-dimensional functions. It is no problem to process such 4-dimensional functions. We will discuss below the computing time required for 4-dimensional processing with our small PDP 11/34 computer.

Step 2. Calculation of the ensemble average triple correlation $\langle I_n^{(3)}(\mathbf{x}, \mathbf{x}') \rangle$ or the ensemble average bispectrum $\langle \tilde{I}^{(3)}(\mathbf{u}, \mathbf{v}) \rangle$ by averaging over all $I_n(\mathbf{x})$:

$$\langle \tilde{I}^{(3)}(\mathbf{u}, \mathbf{v}) \rangle = \frac{1}{N} \sum_{n=1}^{N} \tilde{I}_n(\mathbf{u}) \tilde{I}_n(\mathbf{v}) \tilde{I}_n(-\mathbf{u} - \mathbf{v}). \tag{4}$$

In the case of photon noise, a photon noise bias in the ensemble average bispectrum has to be compensated (Wirnitzer, 1985).

Step 3. Inserting $\tilde{I}_n(\mathbf{u}) = \tilde{O}(\mathbf{u}) \tilde{P}_n(\mathbf{u})$ into Eqn (3) yields for the average bispectrum

$$\langle \tilde{I}_n^{(3)}(\mathbf{u}, \mathbf{v}) \rangle = \tilde{O}(\mathbf{u}) \tilde{O}(\mathbf{v}) \tilde{O}(-\mathbf{u} - \mathbf{v}) \langle \tilde{P}_n(\mathbf{u}) \tilde{P}_n(\mathbf{v}) \tilde{P}_n(-\mathbf{u} - \mathbf{v}) \rangle, \tag{5}$$

where $\tilde{O}(\mathbf{u}) \tilde{O}(\mathbf{v}) \tilde{O}(-\mathbf{u} - \mathbf{v})$ is the object bispectrum and $\langle \tilde{P}_n(\mathbf{u}) \tilde{P}_n(\mathbf{v}) \tilde{P}_n (-\mathbf{u} - \mathbf{v}) \rangle$ is called the speckle masking transfer function. From Eqn (5) the object bispectrum

$$\tilde{O}^{(3)}(\mathbf{u}, \mathbf{v}) = \tilde{O}(\mathbf{u}) \tilde{O}(\mathbf{v}) \tilde{O}(-\mathbf{u} - \mathbf{v}) \tag{6}$$

can be derived since the speckle masking transfer function is greater than zero up to the diffraction cut-off frequency. The speckle masking transfer function is measured in a similar way as the speckle interferometry transfer function, i.e., by observing a calibration star after the observation of the object.

Step 4. From the object bispectrum $\tilde{O}^{(3)}(\mathbf{u}, \mathbf{v})$ the modulus *and the phase* of the object Fourier transform $\tilde{O}(\mathbf{u})$ can be derived. For simplicity, we will use a one-dimensional description in the text below. *Discrete coordinates p, q and r $(r := p + q)$* will be used since discrete coordinates are required for the digital recursive phase reconstruction. For the phase φ_r (phase at coordinate r) of the object Fourier transform $\tilde{O}_r$, we obtained the recursive algorithm

$$\exp(\mathrm{i}\varphi_r) = \mathrm{const.} \sum_{0 < q \leqslant r/2,} \exp[\mathrm{i}(\varphi_q + \varphi_{r-q} - \beta_{r-q,q})];$$

$$\varphi_0 = \varphi_1 = 0; \quad r = 2, 3, \ldots, \quad \text{e.g. 48}; \quad p, q \geqslant 0,$$

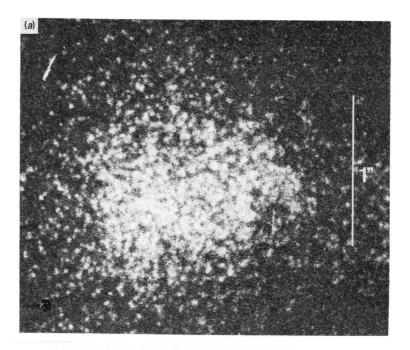

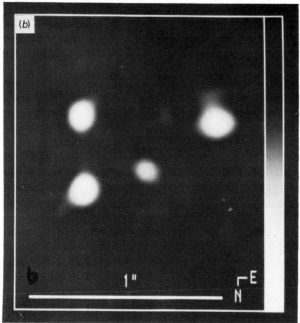

Fig. 10.1. Speckle masking observation of the central object in the giant HII region NGC 3603: (a) One of the 300 reduced speckle interferograms (43 seconds observing time, 1.5-m Danish telescope, La Silla). (b) Diffraction-limited image reconstructed by speckle masking (V-magnitudes: 11.7^m, 11.7^m, 11.7^m, 12.2^m; separations: 0.34, 0.37, and 0.78 arcsec), $\lambda \sim 650$ nm (from Hofmann & Weigelt, 1986a)

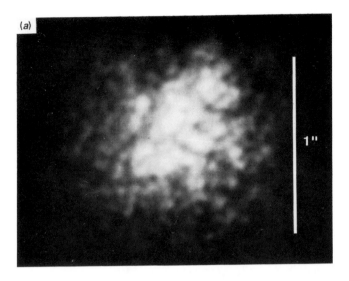

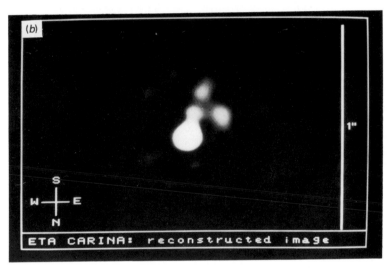

Fig. 10.2. First reconstruction of a true image of Eta Carinae: (*a*) One of the 300 reduced speckle interferograms of Eta Carinae (2.2-m telescope, La Silla). (*b*) Diffraction-limited image reconstructed by speckle masking ($\lambda \sim 850$ nm). The three fainter components are five times fainter than the bright component. The separations are 0.09, 0.18, and 0.22 arcsec (from Hofmann & Weigelt, 1987).

where φ_r is the phase of the object Fourier transform at coordinate r, $\beta_{r-q,q}(=\beta_{p,q})$ is the phase of the object bispectrum, and p, q and r the coordinates of the sampled versions of the object Fourier transform and the object bispectrum. From the obtained object Fourier transform the diffraction-limited image of the object is obtained by an inverse Fourier transformation. Image degradation by the atmosphere and by telescope aberrations are overcome completely. An important advantage of the described phase algorithm is the fact that no phase transfer function has to be compensated.

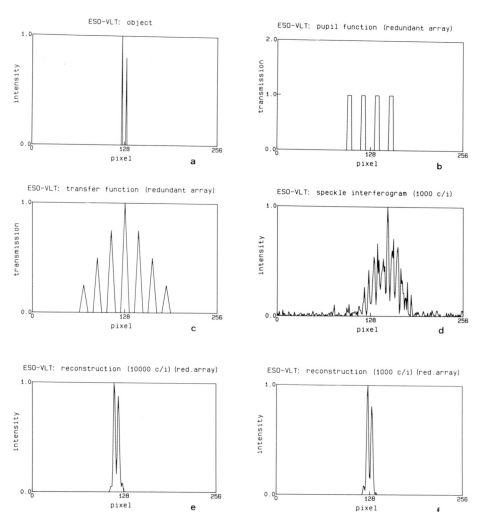

Fig. 10.3. Image reconstruction from optical long-baseline interferograms by speckle masking; simulation of a 25 m–25 m–25 m redundant array of four 8-m telescopes (ESO VLT): (*a*) object; (*b*) redundant pupil function; (*c*) transfer function (with gaps in uv-space); (*d*) one of the 4000 simulated long-baseline speckle interferograms caused by simulated atmospheric turbulence; (*e*) image reconstructed by speckle masking from 4000 interferograms and for *photon noise* corresponding to a mean count number of 10 000 photon counts per long-baseline interferogram; (*f*) as (*e*), but 1000 counts/interferogram (from Reinheimer & Weigelt, 1987).

Fig. 10.4. An artist's impression of a large imaging speckle interferometer in space (ISIS).

In the NGC3603 experiment described below we calculated only 5% of the total bispectrum consisting of 48^4 pixels (the fraction near the axes of the bispectrum which has highest s/n). This small fraction was sufficient for image reconstruction. In this case (5% of the 48^4-pixel bispectrum) the computing time for the reduction of one speckle interferogram is about 7 min for our PDP 11/34 computer.

Figures 10.1 and 10.2 are two speckle masking examples. Figure 10.1 shows one speckle interferogram and the reconstructed image of the central object in the giant HII region NGC3603 (Hofmann & Weigelt, 1986a). Figure 10.2 shows one speckle interferogram and the reconstructed image of Eta Carinae (Hofmann & Weigelt, 1988).

Speckle spectroscopy is a speckle method which yields diffraction – limited objective prism spectra (Weigelt, 1981; Stork & Weigelt, 1984; Weigelt *et al.*, 1986). The raw data for speckle spectroscopy are so-called spectrum speckle interferograms in which each speckle is dispersed by a spectrograph. The reconstruction of the objective prism spectrum is performed by speckle masking.

3 Optical long-baseline interferometry and aperture synthesis

Speckle masking experiments with interferograms recorded by a compact array were described by Hofmann & Weigelt (1986b). In the text below we show that speckle masking can even reconstruct images if the distance between the telescopes is so large that the optical transfer function (autocorrelation of the pupil function) contains many gaps. We have performed digital speckle-masking experiments with different types of pupil functions. Figure 10.3(*a–f*) show an experiment with a simulated redundant array of four 8-m telescopes with telescope distances of 25 m–25 m–25 m (ESO VLT; see Woltjer, these proceedings). It was not necessary to modify the speckle masking algorithm for this case of a coherent telescope array. Similar experiments with a non-redundant four-telescope array and with a ten-telescope array were described by Reinheimer & Weigelt (1987).

Optical long-baseline interferometry and aperture synthesis can be performed on earth or in space (Labeyrie, 1976; Labeyrie *et al.*, 1984; Hofmann & Weigelt, 1986b, Fig. 10.4). In space it will be possible to construct coherent telescope arrays consisting of many independent free-flying telescopes, as proposed by Labeyrie *et al.* (1984). Baselines up to about 100 km seem to be possible. At $\lambda \sim 200$ nm a 40-km array would yield the fantastic resolution of 10^{-6} arcsec!

Acknowledgements

We thank ESO for observing time and the German Science Foundation for financial support of our speckle project.

References

Dainty, J.C. (1985). In *Laser Speckle and Related Phenomena*, Vol. 9 of *Topics in Applied Physics*, 2nd edn, J.C. Dainty, ed. (Springer, Berlin, 1985).

Hofmann, K.-H. & Weigelt, G. (1986a). *Astron. Astrophys.*, **167**, L15.

Hofmann, K.-H. & Weigelt, G. (1986b). *J. Opt. Soc. Am.*, **A3**, 1908.

Hofmann, K.-H. & Weigelt, G. (1988). *Astron. Astrophys.*, **203**, L21.

Labeyrie, A. (1970). *Astron. Astrophys.*, **6**, 85.

Labeyrie, A. (1976). *La Recherche*, **67**, 421.

Labeyrie, A., Authier, B., Biot, J.L., de Graauw, T., Kibblewhite, E., Koechlin, L., Rabout, P. & Weigelt, G. (1984). *Bull. Astron. Soc.*, **16**, 828.

Lohmann, A., Weigelt, G. & Wirnitzer, B. (1983). *Appl. Opt.*, **22**, 4028.

Reinheimer, T. & Weigelt, G. (1987). *Astron. Astrophys.*, **176**, L17.

Stork, W. & Weigelt, G. (1984). 'Speckle spectroscopy', in Proc. ICO-13 Conf. (Sapporo), p. 642.

Weigelt, G. (1977). *Opt. Commun.*, **21**, 55.

Weigelt, G. (1981). 'Speckle interferometry, speckle holography, speckle spectroscopy and reconstruction of high-resolution images from ST data', in ESO Proc. *Scientific Importance of High Angular Resolution at IR and Optical Wavelength*, M.H. Ulrich & K. Kjaer eds (ESO, Garching, Germany), p. 95.

Weigelt, G. & Wirnitzer, B. (1983). *Opt. Lett.*, **8**, 389.

Weigelt, G., Baier, G., Ebersberger, J., Fleischmann, F., Hofmann, K.-H. & Ladebeck, R. (1986). *Opt. Eng.*, **25**, 706.

Wirnitzer, B. (1985). *J. Opt. Soc. Am.*, **A2**, 14.

11

Optical aperture synthesis

J.E. BALDWIN

Summary

First and future applications of the aperture synthesis technique to optical imaging are described.

Aperture synthesis is only one possible technique among many for high resolution optical imaging. So why should we bother to look at yet one more method? I think the technique is attractive because we know it works and we know a lot about it. It has been spectacularly successful at radio wavelengths in providing high resolution images with good dynamic ranges over a wide range of wavelengths. By now there is more than 30 years' experience of using it, thinking about it and implementing it in hardware and software. If the radio and optical analogies are valid, then it must be a way of achieving high resolution optical images. And some of us, reviving the arrogance of youth, would say it is probably the best way.

At radio wavelengths the two essential and complementary techniques are well known to be:

(i) *Aperture synthesis* using the relation that the distribution of brightness across an incoherent source is the Fourier Transform of the spatial coherence function (i.e. of the complex visibilities of the fringes from many baselines in a Michelson stellar interferometer).

(ii) *Closure phase.* A closure phase is derived from simultaneous measurements of a triangle of the three baselines formed by three apertures 1, 2 and 3. In ideal conditions with no atmosphere the observed phases on the three baselines would be Φ_{12}, Φ_{23} and Φ_{31}. These phases are properties only of the source and its

123

position in the sky relative to the interferometer. In practice the phases are affected by the changing atmospheric phase paths above each aperture giving phases $\alpha_1, \alpha_2, \alpha_3$. The observed phases will be $(\Phi_{12} + \alpha_1 - \alpha_2)$, $(\Phi_{23} + \alpha_2 - \alpha_3)$, $(\Phi_{31} + \alpha_3 - \alpha_1)$. Each contains wholly unknown quantities, but the sum of the three, the closure phase $\Phi_{123} = \Phi_{12} + \Phi_{23} + \Phi_{31}$ is independent of the atmosphere.

The methods should carry over directly to the optical regime. The only doubts arise when there are qualitative differences between the radio and optical cases. They are of three kinds:

(i) differences of origin of the atmospheric phase fluctuations;
(ii) lack of satisfactory amplifiers at optical wavelengths;
(iii) the possibility of photon counting at optical wavelengths.

In the radio regime, closure phase is applied in a straightforward way to observations at cm wavelengths in VLBI, and with instruments such as the VLA, where the angular size of the isoplanatic region is larger than the diameter of the primary beam (the Airy pattern) of the individual telescope in the array, so that there is a single unknown atmospheric phase path to all parts of the field of view. Under these circumstances the application of closure phase measurement provides a very powerful method in obtaining images of high quality (see Readhead & Pearson, 1984, for a review).

The phase fluctuations at radio wavelengths arise mainly from variations in the water vapour content and from ionospheric irregularities. They differ in their wavelength dependence, power spectrum and timescales from those at optical wavelengths which are due to rapidly drifting density fluctuations in the lower atmosphere. However, for a small (~ 0.1 m) optical telescope, the size of the isoplanatic patch (~ 20 arcsec) is much larger than that of the Airy disk (~ 1 arcsec) so that the conditions necessary for using closure phase also apply in this case.

At radio wavelengths one may usually ignore the atmospheric variations in phase of the incident wavefront across the diameter of a large paraboloid (they are just of importance at a wavelength of 10 mm for a 100 m paraboloid). At optical wavelengths the phase variations are large and give rise to the speckle patterns, rather than diffraction limited images, obtained with large telescopes. It is essential for interferometry using closure phase that these variations should be small and hence that the diameters of the mirrors should also be small. In practice the largest telescopes that can be used without serious degradation have diameters of $(2 - 2.5)r_0$ where r_0 is

Fried's parameter, the spatial separation at which the rms phase variation is 2.6 radian. A substantial contribution to the phase variations corresponds to a mean phase tilt across the aperture which is corrected for by using autoguiding.

The lack of useful optical amplifiers (which Radhakrishnan has reminded me is a matter of principle and not technological accident) affects the limiting sensitivity of an array of n telescopes but does not prevent it working. The usual condition for the closure phase method to work is that, for each of the three baselines making up the triangle, the signal-to-noise ratio obtained during a single integration time must be $\geqslant 1$. If this is achieved at radio wavelengths, then no penalty ensues if the number of elements in the array is increased. A high gain amplifier at each antenna ensures that the signal for that element can be split into many parallel outputs without loss of signal-to-noise. A large number n of elements is desirable for other reasons; the number of simultaneous baselines is $n(n-1)/2$, the number of independent closure phases is $(n-1)(n-2)/2$ giving a fraction of the total phase information $(1-2/n)$, and the sensitivity in the final image improves as n.

At optical wavelengths there are no coherent broadband amplifiers. Thus the signal from each mirror must be split $(n-1)$ ways to enable all the interferometer baselines to be correlated and the limiting sensitivity, set by photon counting statistics in the maximum usable integration time, is reduced by a factor $(n-1)$. It is a matter of judgement to balance the conflicting advantages of large or small n.

The possibility of counting optical photons is important since the limiting sensitivity is likely to be set by the statistics of the detected photons. However, as Chris Dainty has emphasized in his talk, the quantum nature of light does not undermine in any way the basic relationships in coherence.

So the overall conclusion is that none of the three main differences between the radio and optical wavebands is fatal to the operation of optical aperture synthesis. Its principles are well-established; their experimental verification in a simple arrangement is an important next step. The use of aperture masks on existing large optical telescopes provides a particularly convenient arrangement. The principal technical requirement of maintaining white light fringes in interferometry is easily accommodated by the accuracy of the telescope optics. A group of us, Chris Haniff, Craig Mackay, Peter Warner and myself have therefore carried out experiments using the 88-in University of Hawaii telescope and the Isaac Newton Telescope in order to make

(i) the first measurements of optical closure phases (Baldwin *et al.*, 1986);

(ii) the first optical images using aperture synthesis and closure phases.

In the first experiments in December 1984, using a CCD at 677 nm, masks at a reimaged pupil comprised three apertures with effective diameters of 13 cm. The image of a star shows the Airy disk appropriate to the size of these apertures crossed by three sets of fringes corresponding to the three baselines (Fig. 11.1). The closure phases derived from Fourier transforms of these images were obtained for a range of exposures, as shown in Fig. 11.2. Exposures shorter than 50 msec show high fringe visibilities and closure phases with a small scatter about the value of 0 deg expected for an unresolved star. With exposures longer than the coherence time of the atmosphere across these apertures, the visibilities decreased and the closure phases are randomly distributed through ± 180 deg. Similar results have since been

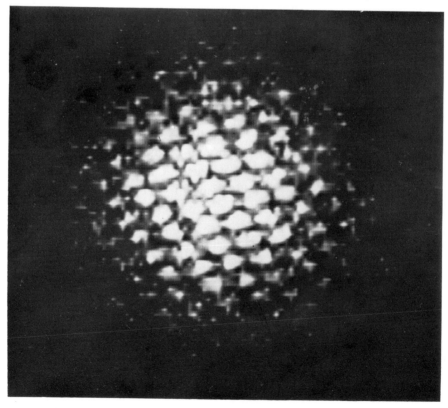

Fig. 11.1. Image of β Ori obtained with a three-hole aperture mask. Exposure 50 msec.

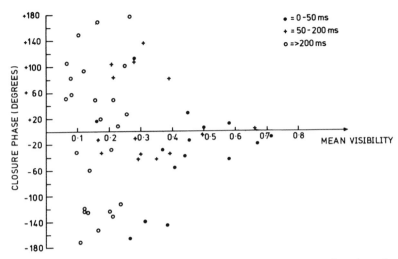

Fig. 11.2. Closure phases from individual exposures as a function of mean fringe visibility and exposure time.

obtained by Readhead, Pearson, Neugebauer & Oke with a CCD on the Hale telescope. The results Tony Readhead has shown me at this meeting indicate that the rms scatter on the closure phases from individual frames increases with exposure and fits that predicted for a Fried model of the atmospheric turbulence. There have also been attempts at closure phase measurement by Robertson, Bailey, O'Sullivan & Frater on the AAT but so far the fringe visibilities have been unaccountably low. I have also learned at this meeting that the first trials of optical closure phase took place long ago. Roger Jennison (1961) outlines a method for obtaining optical closure phases using aperture masks on a telescope. What is new to us is that he tried it out at that time in the laboratory with a small telescope.

The next set of observations we made was in November 1985 using the Issac Newton Telescope and was designed for a first test of the imaging capabilities of the technique for simple objects at low photon rates. A linear arrangement of four 56 mm apertures was used giving a non-redundant array of 6 interferometer baselines at uniform intervals in baseline up to 1 m (Fig. 11.3). The parallel sets of fringes crossing the Airy disk were imaged on the IPCS operating in the photon tagging mode. The photon rates for a + 4.0 mag star, λ Peg, at $\lambda = 512$ nm were only 40 in the exposure time of 17 msec. The visibilities and exact separations of the fringes were obtained from an average of the autocorrelation functions of the individual exposures (Fig. 11.4). Its Fourier transform (Fig. 11.5) shows peaks for the six baselines against a noise background due to the Poisson statistics of the photons. The

Fig. 11.3. 4-hole non-redundant aperture mask used at a 9 mm pupil of the 2.5 m Isaac Newton Telescope. The largest baseline corresponds to a separation of 96 cm.

visibilities are high ($\sim 65\%$) at short baselines and fall at longer baselines due to a combination of instrumental and atmospheric effects. Closure phases for each exposure were then calculated giving a histogram of values (Fig. 11.6) with a wide scatter but showing a peak close to 0 deg. The mean closure phase was calculated from the individual values Φ_i by $\langle\Phi\rangle = \arctan$ ($\sum \sin\Phi_i / \sum \cos\Phi_i$). Notice that this formulation does not require that the signal-to-noise is > 1 in a single exposure, so there is no sharp limit to the instrumental sensitivity. The same procedure would be equally valid at radio wavelengths. The accuracy of the closure phase derived from 2500 exposures

Fig. 11.4. Sum of autocorrelation functions of several thousand individual exposures.

was about ± 5 deg. Observations were repeated at 9 position angles separated by 20 deg by rotation of the aperture mask and detector simulating the configurations attained at radio wavelengths in earth-rotation aperture synthesis. The image reconstructed from the visibilities and closure phases (Fig. 11.7) has a resolution of 70 milliarcsec and is a close approximation to the ideal point response function of the 54 baselines. After CLEANing, the dynamic range in the image in $\sim 100{:}1$. Similar observations of a double star ($\emptyset$ And. $+4.5$, $+5.7$ mag, separation 0.45 arcsec) were reduced by the same techniques and an image reconstructed using the Maximum Entropy Method (Fig. 11.8). The double is well-resolved with a resolution of 30 milliarcsec on the brighter component and 50 milliarcsec for the other. A faint spurious component occurs at the symmetric position in the image. This is due to the very low signal-to-noise in the data; some of the frames had as few as 25 photons and the visibilities were again low due to too-long exposure times, poor focussing, atmospheric tilts and telescope motion. The success of these first images obtained in optical aperture

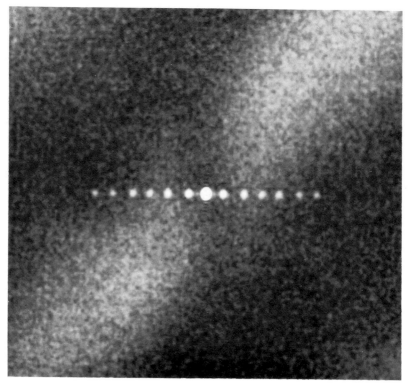

Fig. 11.5. Fourier transform of Figure 11.4 showing peaks corresponding to the six baselines.

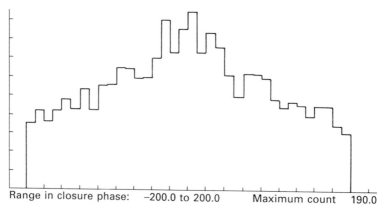

Range in closure phase: −200.0 to 200.0   Maximum count   190.0

Fig. 11.6. Histogram of closure phases from 2500 exposures of λ Peg.

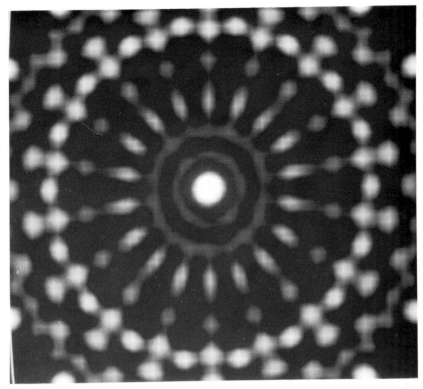

Fig. 11.7. Image of λ Peg obtained using 54 interferometer baselines. The resolution (FWHM) of the central response is 70 milliarcsec.

synthesis is very encouraging for two reasons; first, the technique works well even at low photon rates. The histograms of closure phase values are very similar to the theoretical probability distributions based on the Poisson statistics of the photons in each frame. Secondly, the reductions and imaging were achieved using relatively small modifications of existing radio astronomical software.

In future observing runs at La Palma we plan imaging of more complex objects and investigation of dispersion techniques to obtain larger bandwidths and hence better sensitivity. We shall also use the 36 in telescope at Cambridge for testing detectors and beam combining optics and determining atmospheric parameters for direct application to a more ambitious proposal, a separated element aperture synthesis telescope.

The Cambridge optical aperture synthesis telescope (COAST) is a joint proposal from Mackay and Willstrop at the Institute of Astronomy and

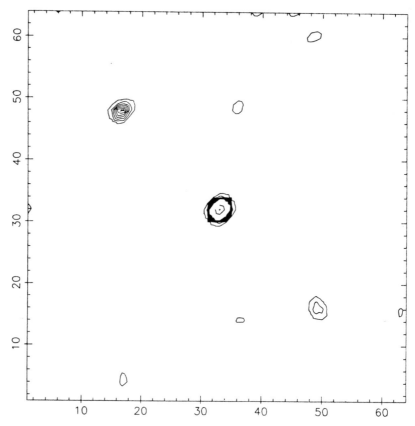

Fig. 11.8. Image of ϕ And obtained using a Maximum Entropy algorithm on the fringe visibility and closure phase data.

myself, Boysen, Scheuer, Warner and Wilson at MRAO. Its principal features and its hoped-for performance are summarized below.

Cambridge optical aperture synthesis telescope

4 Telescopes
Each is a 50 cm siderostat + Cassegrain optics
Baselines up to 100 m
6 baselines, 3 closure phases
Autoguiding for tilt correction of wavefronts
4 paths compensators
Operational wavelength 800 nm – 1.6 μm

Detectors: avalanche photo-diodes, photon counting mode
Real-time reduction
Limiting magnitude (Cambridge) $+10.5$ at $\lambda 800$ nm
$\qquad\qquad\qquad\qquad\qquad +14 \quad$ at $\lambda 1.6 \,\mu$m

The important features of the design of the telescope are the long wavelengths chosen for operation and the use of avalanche photodiodes as detectors. Many factors make it easier to implement optical aperture synthesis at longer wavelengths including the relaxed mechanical tolerances on moving parts and mirror surfaces, the slower timescales of the phase variations and in particular the greatly enhanced limiting sensitivity which can be achieved.

It is generally believed that the spatial spectrum of the phase fluctuations corresponds to a Kolmogorov turbulent spectrum giving an rms phase difference between two points a distance r apart of

$$\Delta\Phi \propto r^{5/6}\lambda^{-1},$$

ignoring the dispersion of air. It must be said that this belief is based on rather few actual measurements and that the limits of its applicability (i.e. the inner and outer scales of turbulence) are unknown. Then, for a given value of the rms phase deviations which is judged to be acceptable, the maximum diameter of mirror which can be used $d \propto \lambda^{6/5}$ and its area $\propto \lambda^{12/5}$. If the timescale of the fluctuations is determined by a wind speed, the maximum integration time for a given degree of blurring in phase is also $\propto d \propto \lambda^{6/5}$. Many optical arrangements, including those which we plan to use, allow for use of a constant fractional bandwidth for a given degradation in performance, i.e. $\Delta\lambda \propto \lambda$. Thus the overall sensitivity of the telescope, measured by photons collected in an integration time, is $\propto \lambda^{12/5}\lambda^{6/5}\lambda^{5/5}$, that is $\lambda^{4.6}$. Because of this very strong factor, the optimum wavelength for maximum sensitivity is the longest wavelength at which photon-counting detectors are available. At present this is at ~ 800 nm but very shortly will be extended to $1.6\,\mu$m.

Avalanche photodiodes are particularly attractive as detectors since they can be operated in photon counting mode and they have very good quantum efficiencies ($\sim 50\%$).

With these characteristics for the telescope, Cambridge becomes a satisfactory site. Darkness of the sky is not a problem for this type of work and even with very conservative estimates of the seeing (2.5 arcsec at 500 nm) the limiting magnitude is expected to lie in the range $+10$ to $+14$ depending on wavelength. The amount of data which such an instrument generates is so large that cloudy nights, of which we do have some, will be welcome. That very briefly is what we want to set our hands to build. Whether we shall be successful in getting the £6 × 10^5 to build it, is not in our hands.

References

Baldwin, J.E., Haniff, C.A., Mackay, C.D. and Warner, P.J. (1986). *Nature*, **320**, 595.

Jennison, R. (1961). *Proc. Phys. Soc.*, **78**, 596.

Readhead, A.C.S. & Pearson, T.J. (1984). *Ann. Rev. Astr. Astrophys.*, **22**, 97.

12

High resolution imaging on La Palma

J.E. NOORDAM

Summary

The combination of modern detectors, development of high-resolution techniques, and an excellent optical/infrared observing site provides the opportunity for major advances in imaging which must be exploited.

The time has come...

The time has come for optical/infrared high resolution imaging from the ground. For the first time since the invention of the telescope, more than 300 years ago, there are realistic prospects of increasing the spatial resolution to the diffraction limit of the telescope. The reasons for thinking this are threefold:

(1) We now have imaging photon counting detectors that allow us to 'freeze' atmospheric turbulence by taking many short exposures, even of quite faint objects.
(2) We now have fast, cheap processors with large memories that allow us to process these short exposures, perhaps even on-line.
(3) Radio aperture synthesis is now a mature science, which has produced many spectacular images. This gives us the mental framework within which to develop its infinitely more difficult counterpart in the optical and infrared.

I hasten to add that the third point in no way dismisses the very valuable (and in many ways heroic) work of the well-known pioneers in optical interferometry, from Michelson to Labeyrie and Weigelt. It merely states that the phenomenal success of radio aperture synthesis gives us confidence and direction in a field where nothing is easy.

Ultimately, of course, optical/infrared high resolution imaging of very faint objects (the most interesting ones) will have to be done from space. Atmospheric turbulence places too much of a limit on integration time and/or the availability of sufficiently bright reference objects within the isoplanatic patch. But before we can ask for the vast sums of money required for a large space interferometer, we will have to demonstrate our techniques from the ground. A natural first step is to make use of existing large telescopes to make diffraction-limited images. The experience gained can then guide the design and operation of a multi-element aperture synthesis instrument.

...and La Palma is a good place for it

The La Palma observatory is more suitable than most for the development and use of high resolution imaging techniques, for the following reasons:

(1) The site has excellent seeing, especially when Sahara dust forms a 'blanket' of constant temperature which very much reduces dome seeing.
(2) The right kind of facilities are being provided to make optimum use of these seeing conditions by effective time-sharing.
(3) The Anglo-Dutch community that is served by the observatory has a strong radio synthesis tradition.

A vigorous program of high resolution imaging, which involves many groups in the UK, the Netherlands, Spain, Ireland and even outside is already in progress, and the first results are about to be published. By offering unique opportunities to a great variety of creative people with different backgrounds, La Palma has the potential to become an important 'catalyst' in the development of high resolution imaging and its many related fields.

Diffraction-limited imaging: we know what to do...

The fundamental limit on the resolution of any telescope is determined by its diameter in wavelengths. At 0.5 μm, the limit is 0.04 arcsec for the 2.5 m Isaac Newton Telescope, and 0.025 arcsec for the 4.2 m William Herschel Telescope. However, the image quality of large telescopes is degraded by atmospheric turbulence, which causes long exposure images to be convolved with a 'seeing disk' of about 1 arcsec.

Various methods have been proposed to obtain diffraction-limited images through the atmosphere. They all follow the same basic recipe:

(1) Take many short (10–50 msec) exposures for which the seeing is 'frozen'.
(2) For each exposure, derive a quantity which can be integrated. Examples of such quantities are the visibility amplitude (or even intensity), the closure phase and the phase derivative.
(3) Integrate this derived quantity over many exposures until the desired s/n ratio is reached.
(4) Reconstruct the image, unambiguously if possible.

...but we do not know yet how to do it

Although there is a fair bit of agreement about what must be done, it is less clear how it can be done most effectively, and how reliable the different techniques are. There is an especially large jargon-gap between optical and radio interferometrists.

A case in point is the speckle masking method of Dr Weigelt: In many ways this is an ideal technique since it uses the entire telescope aperture and thus is suitable for objects as faint as 19th magnitude. Radio synthesists are trying hard to understand how it works in terms of phase closure radio aperture synthesis, especially how exactly it gets round the problem of pupil redundancy. There can be little doubt that the link between the two fields will lead to a much better understanding of the properties of the resulting images.

Other problem areas are background noise, sensitivity and dynamic range of detectors, and extension into spectroscopy.

Image sharpening ($0''2$–$0''5$)

There is also an intermediate technique, which does not quite achieve diffraction-limited resolution but is easier to implement and can be applied to a larger part of the sky. The recipe is as follows:

(1) Subdivide the telescope aperture into areas comparable to the atmospheric seeing cell size (10–50 cm).

(2) For each subaperture, take a large number of short exposures (10–50 msec), to freeze the seeing variations.

(3) Select the 'sharpest' exposures by means of some sharpness criterion.

(4) Correct the selected exposures for image wander and add them all together.

Just like with diffraction-limited techniques, the observed object or a nearby reference must be bright enough to yield sufficient photons per exposure for decision making. But since less information per exposure is required, the reference object can both be fainter (15 mag) and further away (a few arcmin) from the observed object. The achievable resolution (0.2–0.5) is determined by the size of the seeing cells instead of the telescope diameter. Thus, at the cost of resolution, more objects can be observed by this technique.

At present we only do post-processing image sharpening on La Palma, which means that we record the position of all the incoming photons, and reconstruct the sharpened image afterwards. It is also possible to build a feed-through device with fast shutters and tilt-mirrors to feed a beam of 'sharpened' light to other instruments like a charge coupled device CCD or a spectrograph. The problem then is that one has to predict the seeing behaviour for the next few milliseconds. Such devices have been built on Hawaii, and improved versions are under construction in Durham and Roden (NL). The ultimate goal is to have it as a standard option on the acquisition and guide (A + G) box of every telescope.

We now have experience with two modes of Images Sharpening on La Palma, 1D and 2D. In the latter, 2D images are produced with the help of special magnifying optics in the TAURUS box. In the 1D mode the spatial resolution along the slit of the spectrograph is improved in order to separate the spectra of objects that are very close together. Both are implemented on the INT, and both use the IPCS in photon-tagging mode and virtually the same software.

Figure 12.1 shows an example of 2D Images Sharpening. The double star (POU 3018) has a separation of about 5 arcsec. Only the brighter object has been sharpened, and the fact that the fainter one profits also proves that we are doing something right. However, much of the improvement in resolution (from 0.82 to 0.34 arcsec FWHM) is due only to compensation of telescope movement, and not to any selection of sharp exposures. The brighter object is 12th magnitude, but had to be filtered by 0.6 dex of neutral density to

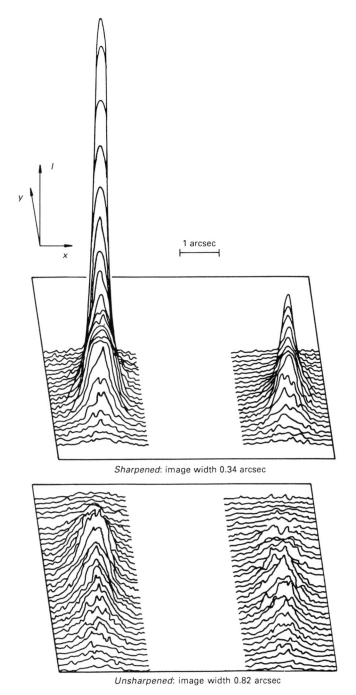

1 arcsec

Sharpened: image width 0.34 arcsec

Unsharpened: image width 0.82 arcsec

Fig. 12.1. An example of 2D Image Sharpening, using the IPCS detector on the 2.5 m INT at La Palma. The double star POU3018 has a separation of about 5 arcsec, and the brighter component is magnitude 12.

protect the IPCS. The result is that the number of photons per short exposure may be too small to decide which are the 'sharper' ones. Some more experimentation is required, preferably under good conditions.

The 'right sort' of facilities

Sir Bernard Lovell in his introduction told us how Prof. Hanbury Brown and his colleagues at Jodrell Bank could only point their radio telescope by carefully tilting the receiver mast. This was a laborious process in which 18 guy wires had to be loosened and tightened in sequence.

But I cannot help thinking that this was nothing compared with the obstacles encountered in doing high resolution imaging experiments (or any other experiments) on large optical telescopes. When after considerable persuasion PATT has allocated two or three nights to such a program, the spectrograph has to be taken off, and the experimenter has half a day to line up his instrument, recalibrate everything and get his software to work. Invariably, the first night is lost in fiddling, the second is used for observing test objects, and on the third night the seeing has become terrible or it is cloudy. Then the instrument has to be taken off the telescope again and if he is lucky the next opportunity will be in six months time when the whole process has to start again. In the meantime he has nothing to show for his efforts and might easily lose interest.

I would say, that radio astronomers have it easy, and the anecdote only illustrates how spoiled they are. But one might also say, that only because they had the 'right sort' of facilities this branch of astronomy could develop so rapidly. Their telescopes were located near their institutes, within easy reach of backup personnel and, very importantly, students to do the hard work and develop new ideas. Moreover, radio observations are almost independent of weather conditions which makes is possible to plan and automate them. And finally, they can observe and experiment in the daytime, without having to fight sleep and exhaustion all the time.

We cannot hope to create the same kind of conditions for optical high resolution experiments on large groundbased telescopes, but we can at least improve the present situation. It would be a giant step forward if experiments could be set up and fully calibrated at an auxiliary focus of the telescope, with the possibility of quickly changing over when the right (seeing) conditions presented themselves. A month of half an hour observations each night, with time to think about the results in between, is infinitely more suitable for high resolution experiments than two or three full nights. This is precisely the idea behind the Groundbased High Resolution Imaging Laboratory (GHRIL) on a Nasmyth platform of the WHT (see below), and an example of the 'right sort' of facility that is planned for La Palma.

The influence of high resolution imaging on the observatory

It has become clear in recent years, that man-made seeing is one of the dominant factors in image degradation. Often the reputation of a telescope is determined by its dome seeing characteristics. A vigorous high resolution imaging program will automatically serve as a 'watchdog' on dome seeing: Not only will its practitioners quickly complain when something is wrong, but they will often have the means and the knowledge to pinpoint the offending condition.

High resolution imaging facilities on La Palma

1985/6 INT: TAURUS-box with IPCS in photon-tagging mode

(a) 2D image sharpening (J.E. Noordam, A.J. Penny)
(b) Optical aperture Synthesis (J.E. Baldwin *et al.*)
(c) Optical aperture Synthesis (J. Bregman *et al.*) SCASIS

1986 INT: spectrograph with IPCS in photon-tagging mode

(a) ID Image Sharpening along the slit (J. Bregman *et al.*)

1987–INT: imaging box (A. Greenaway)

(a) Lots of room ($120*80*40$ cm) and easy access.
(b) Two optical tables and two instrument areas.
(c) Nine detector mounting points.
(d) 4 cm pupil image with large focal depth.
(e) Output focal ratio up to $f/1000$.
(f) $20''$ field, and probe for larger separations.
(g) Possible time-sharing with $f/8$ CCD direct imaging.

1987–INT/WHT: special software on acquisition TV (D.J. Thorne)

(a) Image sharpening.
(b) Seeing investigations and monitoring.

1988–WHT: GHRIL

(a) Enclosure on Nasmyth platform.
(b) Large optical table (250×135 cm).
(c) Photon-counting detector(s) provided.
(d) Data-acquisition infrastructure provided.
(e) Rapid change-over between foci: Flexible observing!

1990 (?) A multi-element optical aperture synthesis array.
2000 (?) Ready for the real work in space.

13

Large mirrors

J.R.P. ANGEL

Summary

The next generation of optical and infrared telescopes needs to use larger mirrors, with a diameter of around 8 m. These mirrors should ideally be lightweight, rigid and thermally responsive; characteristics that can be realized in a honeycomb sandwich structure that is ventilated. They should also have short focal length, to give improved tracking and pointing accuracy in wind, and to minimize construction costs for the telescope and its enclosure. Primary focal ratios as fast as $f/1$ are desirable and can yield good wide-field imaging at secondary foci, provided the primary figure is approximately paraboloidal. This paper describes the development of method and facilities to cast and polish such mirrors.

Introduction

Mirrors for optical telescopes have been a disaster. They are too small, too heavy, with too much thermal inertia and too long focal length. They have resulted in telescopes that are cumbersome, expensive and generally limited in image sharpness by their own locally generated seeing.

I have been fascinated at this meeting to hear of the early days of radio astronomy, when Hanbury and the other pioneers were developing telescopes from scratch. Strange as it may seem, we are at a phase now in optical astronomy where things are not so different. In order to make a new generation of larger telescopes, we are also forced to go back and work out from first principles how to do it. The conventional technology has simply run out of gas. Radio astronomers at this meeting are looking towards

143

optical interferometry as a new field where their expertise can be of value. I believe also that their ideas on telescope construction would be of great benefit. Optical telescopes should probably look much more like radio telescopes than they do now!

The one radical departure from conventional optical telescopes design has been the MMT (Multi-Mirror Telescope; Beckers *et al.*, 1981). This was built and is operated jointly by the University of Arizona and the Smithsonian Astrophysical Observatory. It consists of six 1.8 m telescopes all mounted in the same telescope 'tube', with all the light relayed to a common focus. It has the light-gathering power of a 4.5 m single dish, but is light, short and stiff.

In the MMT the rationale of using several smaller dishes to make a single telescope was not primarily to achieve resolution, as it is in radio arrays. Rather, it was to avoid the difficulties of fabricating, handling and supporting a single large mirror to the much higher tolerance needed at optical wavelengths. It has worked well, and has proved the practicality of this type of telescope, even for operation as a coherent phased array. A great advantage in construction cost and in tracking accuracy was realized by the compactness of the telescope. However, it became clear that if a very short focus large mirror could be used to get the same compactness, there could be still further advantages in light-gathering power, field of view and in operating costs.

Size, shape and accuracy

How big could single mirrors for optical telescopes be made and how short in focal length? These are questions that a group of us at the University of Arizona, including Nick Woolf, Peter Strittmatter and myself, began thinking about after the MMT development. Size is of course limited by the accuracy needed. The largest steerable radio dishes of up to 100 m in diameter maintain a surface accuracy of around 500 μm rms. For efficient diffraction limited performance the rms surface error must be about $\lambda/30$ (Ruze, 1966). Thus for the shortest wavelength radio window, at 350 μm, this should be $\leqslant 12 \mu$m rms. The upper limit for dishes of this accuracy is about 10 m in diameter. Thermal distortion is a significant factor for these telescopes and led to the choice of low-expansion carbon-fibre material for the Max Planck–Arizona submillimeter dish (Baars & Martin, 1987). For optical telescopes that need still much higher accuracy, the criteria are more complex, and we must take a moment to consider what is needed.

To be diffraction-limited at 300 nm wavelength, the ultraviolet limit of transmission of the atmosphere, a mirror would need to be 1000 times more accurate than for the sub-mm band. In practice, such accuracy is only needed on small spatial scales. Larger errors are acceptable on large scales as long as most of the image remains within the diameter set by atmospheric turbulence limitations. This is about 2×10^{-6} radians at high sites under the best seeing conditions. Light diffracted by errors of spatial scale l goes into a halo of angular size λ/l. It follows that for good ultraviolet performance, the surface should be diffraction limited (< 10 nm rms) up to scales l of 15 cm. On larger scales, the surface criterion is that the telescope wavefront aberrations on each scale should be smaller than those of the atmosphere under the best possible observing conditions. These considerations lead to a goal for telescope aberrations given by the solid line in Fig. 13.1, from Angel (1987). Here is plotted the allowable rms wavefront error (twice the surface error), measured between points separated by l. It increases from ~ 20 nm for l up to 5 cm to 200 nm on a scale l of 1 m.

The largest telescope that meets these exacting criteria is probably the new 4.2 m William Herschel telescope. David Brown (1986) has achieved the required surface smoothness in an $f/2.5$ mirror, using a full-size pitch lap. The global accuracy requirements are met by a thick disc of glass ceramic of negligibly small expansion coefficient. As in all present large optical telescopes, the mirror shape is maintained by its rigidity, the glass being supported at many points in a stress-free condition. Support force errors as the mirror changes orientation are kept low enough so the mirror shape remains within specification.

It is both scientifically desirable and technically feasible to build telescopes with twice the mirror diameter of the Herschel telescope. The size goal we adopted for the next generation of optical telescope mirrors is 8 m (Angel & Woolf, 1983). This is large enough that diffraction in the infrared, 0.3 arcsec in the clear 10 μm atmospheric band, will not usually be limiting compared to atmospheric 'seeing'. On the other hand, a single 8 m piece of glass is not too large to be transported to remote mountain sites.

Our approach to making fast 8 m mirrors is, in a sense, very conservative. We would like to make a glass substrate rigid enough so that properly supported it will hold its own shape. This is not necessarily required. Mirror shape could be continually monitored during operation and servos used to

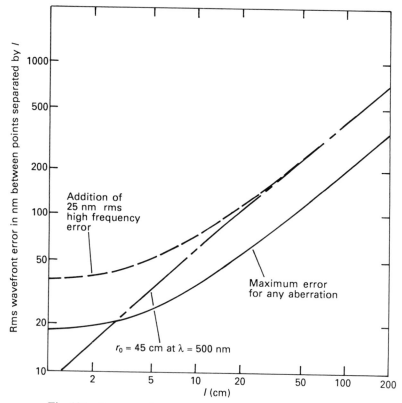

Fig. 13.1. Rms wavefront error measured between points separated by *l*, plotted against *l*. The solid line is the goal for individual aberration terms, derived from conditions of best seeing and diffraction.

correct shape errors, but simple rigidity is better if achievable. People can only remain standing because of continuous servo action of the brain. Statues do it effortlessly for thousands of years.

We would like also to be conservative in mirror polishing. The reason that existing optical telescopes have such long focal lengths ($\geqslant f/2$) is that no one has developed a practical method to polish the strongly aspheric surfaces of a fast paraboloid. The natural smoothing action of a rigid large polishing tool works only on spherical or flat surfaces. On aspheric surfaces the tools have to be either small or flexible, and their smoothing action operates only on a spatial scale over which the surface approximates a sphere to within a fraction of the wavelength of light. For an 8 m f/1 mirror this scale is about 3 cm. Larger rigid tools could, however, be used if their shape were changed by actuators as they move over the mirror. Our development of a technique to do this is described further below.

Mechanical thermal design of glass mirror blanks

A recent discussion of these topics is given by Angel (1987). Here we give a summary of the design factors. An 8 m glass disc, rigid enough to hold the tolerance given above, would have to be about 0.6 m thick. It would then need a hundred or so support points distributed to balance the weight evenly all over, and the applied forces would have to be accurate to about 0.1% in both strength and direction. Discs much thinner than this need so many supports of such high force accuracy that a servo becomes necessary, especially because of the unpredictable forces of wind loading.

A solid glass disc of 60 cm thickness would both be very heavy (70 tons) and have a long thermal time-constant (several days). Thermal inertia is a problem because a mirror warmer than the ambient air sets up convective turbulence, degrading image quality. To eliminate these effects effectively, a time-constant of an hour or less is needed. With this kind of response a mirror follows temperature changes fast enough that it will generally be within 0.25 °C of ambient temperature; the wavefront error from convection is then smaller than the goal of Fig. 13.1.

Dramatic improvement in thermal response can be realized in honeycombed structures that are ventilated. The mass is less, the surface area for cooling is increased, and the time-constant for thermal diffusion to the glass surface is reduced. The optimum form is a honeycomb sandwich, with front and back facesheets separated by a hexagonal or square rib structure. These structures are even slightly stiffer against deflection under their own weight than solid mirrors of the same external dimensions. The specific design we have in mind for 8 m blanks has an average thickness of 60 cm with 80% of the mass eliminated by the creation of internal cavities. The facesheets are 3 cm thick, and the honeycomb cells 20 cm across. For such a structure the ratio of surface area to mass is more than 20 times greater than for a solid mirror. If the natural cooling of the front face by wind and convection is matched by forced air flow on the other surfaces, then a time constant of an hour will result.

Andrew Cheng and I have made laboratory experiments to find the best way to achieve internal ventilation. Each cell can be accessed by a 10 cm hole in the backplate, and will be individually ventilated. A good thermal model for the whole mirror can thus be made by constructing a single cell whose walls are $\frac{1}{2}$ of the proposed thickness, and are well insulated. Schemes that use internal plumbing to draw or blow air into each cell have been described (Wong, 1984; Cheng & Angel, 1986). We have recently found (Cheng, 1987)

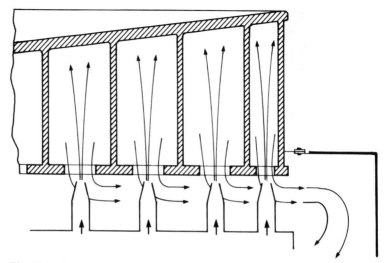

Fig. 13.2. Cross-section showing the method for ventilating a glass honeycomb mirror. Air at a pressure of ~ 2.5 cm of water in the manifold below is directed into each cell by nozzles below the miror.

that it is easier and better to jet air into each cell from below, as shown in Fig. 13.2. The resulting turbulence as the air finds its way back out of the inlet hole results in a high thermal coupling, as well as uniform cooling throughout the cell. We find, empirically, that a flow rate of 5 litre s^{-1} per cell gives the desired one-hour time-constant.

The type and quality of glass to be used in mirror blanks needs some discussion because thermally induced distortion of the mirror substrate can also introduce aberrations. This places constraints on the homogeneity of thermal expansion coefficient and on temperature gradients within the glass. One can show that to keep the reflected wavefront within the tolerance of Fig. 13.1, the products $(T - T_0) \Delta\alpha$ and $\alpha\Delta T$ must both be $< 3 \times 10^{-7}$. Here α and $\Delta\alpha$ are the coefficient of thermal expansion and its rms variation; T and ΔT are the glass temperature and its spatial variation; T_0 is the temperature at which the mirror was figured and tested. The coldest temperatures at the best sites are about 30°C below the figuring temperature, so $\Delta\alpha$ should be $< 10^{-8}$ °C^{-1}. The best quality glasses made in the large quantities needed for a mirror blank can reach this level of homogeneity. Both borosilicate glass and glass ceramics show rms variation $\Delta\alpha$ of about 1×10^{-8} °C^{-1}.

Our choice of glass is restricted more by the internal gradients ΔT that are developed. In thick blanks gradients of > 1°C are inevitable, because of changes in ambient temperature and poor thermal diffusivity. Such blanks must be made of 'zero expansion' glasses with $\alpha < 3 \times 10^{-7}\,°C^{-1}$, such as Cervit and Zerodur (glass ceramics) or titanium silicate (Corning's ULE). Pure fused silica ($\alpha = 5 \times 10^{-7}\,°C^{-1}$) would be marginal in a thick blank.

All of these very low expansion materials would be very difficult to fabricate into the complex structure of an 8 m honeycomb sandwich. The precisely controlled crystal growth in glass ceramics needed to get low and uniform expansion can be achieved only in solid blanks, which would have to be machined out. Silica and ULE are so refractory that structures can be made only by fusing or bonding together plate and rib sections that are previously machined to shape.

A vitreous glass that can be melted to moderately low viscosity and cast in a mold would be much easier to fabricate. No such glass has very low expansion coefficient, but the best of the borosilicates, with $\alpha = 2.8 \times 10^{-6}\,°C^{-1}$ would be satisfactory provided $\Delta T < 0.1\,°C$ can be maintained by ventilation. Our laboratory experiments show that the internal equilibrium achieved with 5 litre s^{-1} per cell has a peak-to-valley range of temperature of 0.1°C, for typical night-time cooling rate of 0.25°C h^{-1}. Ventilation can thus serve two crucial and distinct functions in honeycomb mirrors. It both eliminates mirror seeing and allows the use of borosilicate glass which can be cast into the elaborate structure. A process for doing this was developed in experiments made by John Hill and myself, and is described in the next section.

Casting glass honeycomb mirrors

The Palomar 5 m mirror was cast 50 years ago in a circular mold with a pattern of brick blocks on the floor. Glass poured into the mold filled the spaces between the blocks, forming a triangular rib structure with a facesheet above. Both face and ribs were 10 cm thick. What we are developing at the Mirror Laboratory of Steward Observatory can be thought of as an extension of this method with a more elaborate mold to make a full honeycomb sandwich structure rather than a ribbed plate. Closely spaced hexagonal blocks are anchored to the floor of the mold. Their separation is only 1.3 cm, to form a much lighter honeycomb rib structure. Also the blocks are raised 3 cm off the base, so the glass runs under the blocks to form a fairly continuous back plate, pierced only by holes left by the anchoring bolts of silicon carbide. We don't ladle in liquid glass, but stack the mold full of glass chunks that are melted in place.

As was done for the Palomar blank, the top faces of the blocks are shaped to the desired mirror curvature. But to avoid the expense and time of grinding out a deeply dished mirror face, the furnace is built on a turntable, so the correct curvature can be formed directly during casting by spinning at the appropriate speed. For an 8 m $f/1$ mirror this speed is 8 rpm. Because of uncertainty in shrinkage of the mold blocks during timing we cannot directly cast the correct face sheet thickness of 3 cm. We aim for about 1 cm excess, to be subsequently machined off, along with bubbles or other surface imperfections.

The use of a lightweight honeycomb of quick thermal response has an advantage at the time of manufacture as well as in operation. The time needed to anneal and cool the glass is typically 50 times the time-constant τ of the glass and mold. For the Palomar blank τ was one week, and a year was spent in cooling. We anticipate $\tau <$ a day for 8 m blanks, and thus less than two months in cooling. The uniform thickness faceplate achieved by spin casting is essential for keeping the annealing time short.

The casting method has been developed in a series of experiments to investigate materials and structures. Several modern borosilicate glasses have been tried. The standard material made in large quantities (Corning's 7741, Schott's Tempax) has a coefficient at room temperature of $3.25 \times 10^{-6}\,°C^{-1}$, and can be cast at a temperature of 1150 °C, when the viscosity is 10^5 poise. We prefer to use Ohara's E6, a glass made in one-ton batches in clay pots and with a lower expansion coefficient (2.8×10^{-6}). Its casting temperature is the same, but its liquidus temperature (above which crystals go back into solution) of 950 °C is relatively low, and results in cleaner castings.

A most useful discovery we made early on is that borosilicate glass can be cast against the ceramic fibre materials commonly used for refractory insulation. The fibres are made from alumina-silica glass. There is very little chemical reaction, even when the glass is in contact for many hours at casting temperature. All the parts of the mold are either made of or faced with this ceramic fibre. In particular, the hexagonal cores are vacuum formed from a fibre slurry. When the casting is complete, these cores are removed from the glass honeycomb by means of a high-pressure water jet.

A specially made rotating furnace of 2 m capacity was used in 1985 to make the 1.8 m $f/1$ honeycomb sandwich blank shown in Fig. 13.3. This is 35 cm thick, and took three weeks to anneal and cool. Details are given by Goble, Angel & Hill (1986). Since then we have moved into a new laboratory built for making 8 m castings. The engineering group led by Larry Goble has now

Fig. 13.3. A honeycomb sandwich blank 1.8 m diameter cast in one piece from borosilicate glass. The front face is formed with an $f/1$ curve by spinning the furnace when the glass is liquid.

completed a 12 m diameter turntable, equipped with a 2.2 MW power distribution system with slip rings, and a drive system for up to 12 rpm rotation. This turntable is shown in Fig. 13.4. Construction of a furnace on this turntable is now under way. It is being configured initially for casting intermediate size trials at 3.5–4 m diameter. Two of these are scheduled for 1987. First big castings will be attempted once the intermediate size is mastered, in 1988 or 1989.

Polishing with an actively stressed lap

A number of considerations led to our development of a new polishing method for very fast mirrors. Harland Epps has shown a number of optical solutions for correctors for paraboloidal or hyperboloidal $f/1$ primaries that give wide, flat foci with good image quality. We favor the corrected Cassegrain solution that gives a 40 arcmin flat field with achromatic $\frac{1}{2}$ arcsec images (Epps, 1986). It incorporates prisms for the correction of atmospheric dispersion. Accurate alignment is needed for $f/1$ optics, but as Meinel (1960) pointed out, because short tubes are stiffer,

Fig. 13.4. Construction in progress of the furnace for spin casting 8 m mirrors.

structural bending considerations may actually favor faster focal ratios. Optical testing during fabrication is possible with correctly designed optics and, again, Epps (1986) has designed good correctors for null testing $f/1$ primaries. The secondaries are actually more difficult to test, but the method of a transparent Hindle sphere (Simpson, Oland & Meckel, 1974) used by David Brown to figure secondaries up to 1.5 m diameter, looks viable.

The difficulty lies in the polishing of fast paraboloids, because of their strong asphericity. The radial and tangential curvatures at radius r of a paraboloid of diameter D and focal ratio F are given by

$$C_r = \frac{1}{2DF}\left(1 - \frac{3}{8}\frac{r^2}{D^2 F^2}\right),$$

$$C_t = \frac{1}{2DF}\left(1 - \frac{1}{8}\frac{r^2}{D^2 F^2}\right).$$

Thus the radial curvature at the edge ($r = D/2$) of an $f/1$ mirror is 10% less than that at the centre, and 7% smaller than the tangential curvature at the same point. Existing polishing methods have not produced telescope mirrors with curvature variations of more than 1%. As we discussed earlier, conventional methods would be obliged to use very small or very flexible laps on strongly aspheric surfaces, and these have very poor smoothing action on scales > 3 cm. The polishing of an 8 m $f/1$ surface by these methods would be an extremely tedious and expensive process.

How can the powerful smoothing action of a large, rigid polishing tool be used to advantage when the curvature changes so strongly? One method that has been exploited is to use a slightly flexible glass substrate, and to bend it with the opposite changes in curvature. Then if it is polished as a smooth sphere, when the bending forces are released it will take on the desired shape. This method is used to make thin axisymmetric surfaces such as Schmidt plates. Furthermore, Lubliner & Nelson (1980) have also shown that edge moments applied to a uniform disc with a parabolic surface can induce changes of shape to make it fit off-axis sections of the same parabola. This principle is to be used to make 1.8 m mirror segments of 0.075 m thick discs for the Keck telescope. However, glass thin enough to take the large bending for $f/1$ mirrors would be too flexible to use.

The opposite approach is to take a slightly flexible polishing tool, and to explicitly bend it with the required shape changes as it moves over a rigid mirror surface. For a full-size tool working a parabola the required distortion as the tool moves off centre is coma. In a large mirror the bending moment provided by gravity as the tool overhangs the mirror can be used to induce such bending. David Brown (1986) used tools that are stiffened with concentric rings to inhibit astigmatic bending, and whose thickness at the edge is chosen so the overhang moment induces the required comatic bending for displacements up to 20–30 cm. This method was used to make the AAT, UKIRT and Herschel telescope mirrors. It might be pushed as far as $f/1$, but probably would not yield the required accuracy of bending.

Another way to bend full-size laps is with edge-levers connected by springs to a point on the mirror axis. Then as the lap is stroked away from the centre, the springs lift the overhanging edge and relax the opposite side. This too induces coma, but of the opposite sign to the overhanging lap weight. It is thus suitable for making secondary mirrors. Peter Wizinovich and I have demonstrated this method by grinding and polishing a strong asphere on a 20 cm flat.

For large $f/1$ primary mirrors we are developing a new method in which a subdiameter lap is bent by edge-moment actuators under computer control (Angel, 1984). We envisage that for 8 m mirrors the lap would be about 2 m in diameter and of aluminum some 8 cm thick. Such a lap has the high rigidity for very strong smoothing action, but the bending needed to change the sagittal depth by 5 mm is within the elastic and fatigue limits of the material.

The lap will carry with it a system of motor driven moment actuators that change its shape as it moves. Each of the 24 actuators will consist of a moment arm bolted to the disc edge, whose top end is connected by a piano wire to the opposite edge of the disc. The tension in the wire can be varied by a motor-screw arrangement. Under tension a strong bending moment is produced at the arm, with almost no corresponding moment on the other side. In this way the tensions of the full set of actuators can be adjusted to produce comatic distortion, as well as focus and astigmatism changes needed for sub-diameter laps. In operation all the wires are under high tension when the lap is at the mirror centre, and as it moves out the actuators are relaxed to decrease the curvature, those nearest to the outside edge being relaxed the most.

Such a stiff lap will only polish correctly if the induced shape changes match exactly those needed. We calculate the changes in curvature must be accurate to 0.1–0.2% of its maximum value. At this level we have to check very carefully that the actuator system, which produces compressional forces as well as bending moments, is really capable of making the right deformations. Also the control servo system needs to be very well engineered. Direct measurement of models and finite element analysis show the actuator system will give accuracy to within 1% over the full disc, better if the very edge is not used. Dick Young and Bob Nagel at Steward have developed a prototype actuator with a torque motor, ball screw and lever actuator that promises to meet the 0.1% dynamic control accuracy at 600 lb. force. Buddy Martin and I will shortly test the method by polishing the 1.8 m $f/1$ honeycomb blank shown in Fig. 13.3. The lap will be of aluminum 60 cm in diameter, and 3 cm thick.

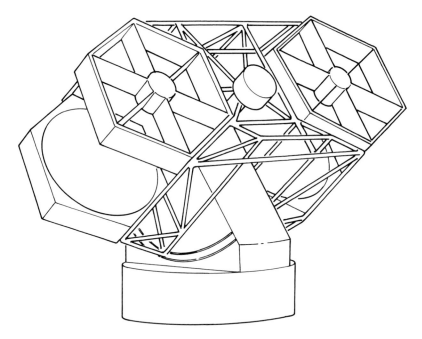

Fig. 13.5. Sketch of a design concept of the Columbus telescope, with two 8 m $f/1$ mirrors in tandem.

The future

If we succeed in our efforts to cast the 8 m blanks, and to polish them at $f/1$, then it will become possible to construct compact 8 m telescopes at a cost comparable to that for present day 4 m telescopes. Plans are quite well advanced to refit the MMT with a 6.5 m $f/1$ mirror. Two consortia have formed to build telescopes with 8 m mirrors. The Universities of Arizona and Chicago, Ohio State University and Italy (Arcetri Observatory) plan a binocular telescope with two 8 m $f/1$ mirrors on Mt Graham, Arizona. A candidate design for this telescope is sketched in Fig. 13.5. The Carnegie Institution with Johns Hopkins University and Arizona plan a single 8 m $f/1$ telescope for Las Campanas in Chile. The National Optical Astronomy Observatory has set the goal of making a four mirror telescope, possibly preceded by a single 8 m telescope.

Two crucial demonstrations of technology are needed before these plans can become fully fledged projects. Trial castings of 3.5 m will be started shortly, and we must make at least one successful blank at this size to prove the method. This blank is destined for the ARC consortium telescope in New Mexico, already under construction. We also must demonstrate the 1.8 m $f/1$ mirror polished to the high quality of Fig. 13.1. Our goal is to do this in 1988.

Acknowledgment

The technology developments described are based on the ideas and hard work of many. I would like to acknowledge particularly the contributions of Steward Observatory Mirror Lab and Technical Support group. Major funding for mirror casting and polishing has come from the National Science Foundation, with casting under subcontract number 2788032 from the National Optical Astronomy Observatories. NASA has supported glass technology and stressed lap polishing development. The University of Arizona has also generously supported this work.

References

Angel, J.R.P. (1984). In *Very Large Telescopes, Their Instrumentation and Programs*, eds Ulrich, M.H. & Kjar, K., ESO, Garching, F.R. Germany, p. 11.

Angel, J.R.P. (1987). *Identification, Optimization and Protection of Optical Telescope Sites*, eds Millis, R., Franz, O., Ables, H. & Dahn, C., Lowell Observatory, Flagstaff.

Angel, J.R.P. & Woolf, N.J. (1983). *Proc. XI Texas Symposium on Relativistic Astrophysics*, ed. Evans, D., *Annals N.Y. Acad. Sci.*, **422**, 163.

Baars, J.W.M. & Martin, R.N. (1987). *Proc. ESA Conference on Space-Borne Sub-Millimeter Astronomy Mission*, Segovia, Spain, p. 247.

Beckers, J., Ulich, B., Shannon, R., Carelton, N., Geary, J., Latham, D., Angel, R., Hoffmann, W., Low, F., Weymann, R. & Woolf, N. (1981). In *Telescopes for the 1980's*, eds Burbidge, G. & Hewitt, A., *Annual Reviews Monograph*, p. 63.

Brown, D.S. (1986). *Optical Fabrication and Testing Technical Digest*, Optical Society of America, Washington, D.C., p. 48.

Cheng, A.Y.S. (1987). Ph.D. Thesis, University of Arizona.

Cheng, A.Y.S. & Angel, J.R.P. (1986). *Proc. S.P.I.E.*, **628**, 536.

Epps, H.W. (1986). MMT Technical Report # 19.

Goble, L., Angel, J.R.P. & Hill, J.M. (1986). *Proc. S.P.I.E.*, 571.

Lubliner, J. & Nelson, J.E. (1980). *Appl. Opt.*, **19**, 2332.

Meinel, A.B. (1960). In *Telescopes*, ed. Kuiper, G.P., & Middlehurst, B.M., p. 25.

Ruze, J. (1966). *Proc. I.E.E.E.*, **54**, 633.

Simpson, F.A., Oland, B.H. & Meckel, J. (1974). *Opt. Eng.*, **13**, p. G101.

Wong, W-Y. (1984). *Proc. S.P.I.E.*, **444**, 211.

14

Wide field telescopes and high redshift QSO surveys

C. HAZARD

Summary

The use of wide field telescopes – Schmidt telescopes in particular – to search for quasars (QSOs) is discussed. It is shown how the search techniques have evolved, and the relative success of the different techniques is considered. These techniques include selection by radio astronomy identifications, broad-band colours, medium-band colours, variability, and grism/objective prism methods. The evolution has been towards broad-band colour selection via automatic machine searches of Schmidt plates, and the details of this process are presented. The success in detecting QSOs of very high redshift is discussed, together with the selection effects. Finally the implication of these searches for future surveys to redshifts in excess of 4 is considered.

1 Introduction

Schmidt telescopes have made major contributions in many areas of astronomy ranging from the measurement of the orbits of asteroids to searches for the most distant objects in the Universe. However, the primary purpose of at least the larger instruments has been in searches for those objects which can be most fruitfully studied in detail with large reflectors. With their wide fields of view, up to 6×6 deg, they are ideal survey instruments and particularly valuable in searches for very rare objects, such as very high redshift QSOs, whose discovery requires surveys over very large areas of sky.

When the identification of 3C273 and the measurement of the redshift of its optical counterpart in 1963 led to the recognition of QSOs as a new class of extra-galactic objects, it opened up a whole new area for Schmidt telescopes and particularly for the Palomar 48-in Schmidt. The main motivation for the Palomar Schmidt was to provide a deep two colour atlas of the sky to allow the selection of faint galaxies for study by the Palomar 5-m reflector. The resulting Palomar Observatory Sky Survey (POSS) was to have applications far beyond those originally envisaged, among the most important of these being its extensive use in radio source identification programmes. The two colour plates were particularly valuable in the early QSO identification programmes when the radio source catalogues had a positional accuracy of no better than about ± 1 arcmin, far too large to allow an identification with a stellar object on the basis of positional agreement alone. However, it was found that even with this limited accuracy reliable identifications could be made with QSOs, provided that they were restricted to stellar objects which appeared blue in the POSS or which were shown by UBV photometry to have an ultraviolet (UV) excess relative to the majority of galactic stars.

The large UV excess, which was an observed characteristic of all the early radio QSOs, also formed the basis of the early searches for radio-quiet QSOs (Sandage, 1965). The earliest optical surveys used a two image technique in U and B or a three image technique in U, B and V with the images being taken on the same plate and the QSOs selected by a simple eye inspection. Later, separate plates in each colour band became more usual. This was a highly successful procedure, although it was soon recognized to be subject to severe redshift dependent selection effects and biassed against QSOs with $z > 2.2$. It was the origin of the present multi-colour search techniques.

Optical variability was another QSO characteristic recognized shortly after their discovery and in principle it can be used to obtain redshift independent samples. However it has not been widely used so that redshift dependent colour selection remained a feature of all large optically selected QSO samples until the introduction of slitless spectroscopy, using either a low dispersion objective prism on a Schmidt telescope or a transmission grating-prism combination on a large reflector. It was used in the early 1960s by Markarian and his colleagues (e.g. Markarian, 1967; Markarian, Lipovetskii and Stepanion, 1979) to search for galaxies with strong UV continuum spectra. Many of these 'Markarian galaxies' were later found to be emission line galaxies, Seyfert galaxies and QSOs but the earliest applications of slitless spectroscopy with a Schmidt telescope for the specific selection of QSOs by their emission lines were by Smith (1975) and Osmer &

Smith (1976) using the 75 cm Curtis Schmidt at Cerro Tololo, Chile. The method is particularly useful for detecting strong emission line QSOs at $z > 1.8$ when the strong Lyα emission line is redshifted above the atmospheric cut-off although weak line lower redshift objects can also be recognized from the colour information in the prism spectra. It is, therefore, a valuable complement to the UBV technique which has its bias against objects with $z > 2.2$. Over the past decade, objective prism surveys have become common-place and have been carried out by various groups on a number of Schmidt telescopes using a variety of photographic emulsions and prism dispersions. The technique, based on eye searches or machine analysis (Clowes *et al.*, 1984; Hewett *et al.*, 1985), is the most powerful available for finding emission line QSOs with the added advantage that the redshifts of many of them can be measured directly from the prism spectra; colour surveys yield only QSO candidates which must be further examined on objective prism plates or studied spectroscopically on large reflectors. It also permits the direct selection of unusual QSOs, e.g. broad absorption line (BAL) QSOs, from details of the prism spectra. The technique has proven particularly powerful with a IIIa-J emulsion and the low dispersion 44 arcmin prism (Nandy *et al.*, 1977) on the 48-in UK Schmidt Telescope (UKST) at Coonabarabran, Australia. A good deep UKST IIIa-J objective prism plate can reach to a continuum magnitude of $\approx +20$ and yield up to 500 QSOs per plate.

The UKST and the similar European Southern Observatory (ESO) Schmidt at La Silla, Chile were constructed to extend the POSS over the Southern Sky for the benefit of the new large reflectors in Chile and Australia. However, by taking advantage of their better optics and of improved emulsions and photographic techniques since the making of the POSS, they have taken survey work to new levels. Using hyper-sensitized plates, the new survey reaches about two magnitudes fainter in the blue (J) and red (R). Furthermore, it is being extended to a third colour in the far red (I) which will permit a useful colour classification from the survey data alone. Both the UKST and the ESO Schmidt are provided with objective prisms and the large collection of UKST prism plates in the plate library at the Royal Observatory, Edinburgh (ROE) is a valuable supplement to the three colour survey. A new Palomar atlas will soon provide similar quality material in three colours in the North but as yet the Palomar Schmidt has no objective prism.

The three-colour survey of the Southern Sky and the UKST prism plates in Edinburgh form an excellent primary data base for large-scale QSO surveys. However, there is no single search procedure that can give an unbiassed sample covering all types of QSOs over all redshifts so that, in general, the survey material needs to be supplemented in selected regions by additional information. This could include (1) additional broad band colours (2) observations using narrow or intermediate band filters (3) plates taken at different epochs for variability studies and (4) deep radio surveys (Hazard, 1980). Until recently it has been impossible to examine the properties of more than a small fraction of the objects on a single deep Schmidt plate using UBV photometry let alone using this multi-technique approach. However, the new generation of measuring machines such as the automatic plate measuring machine (APM; Kibblewhite, *et al.*, 1983) at the Institute of Astronomy, Cambridge, COSMOS at ROE (McGillivray & Stobie, 1985) and the Minnesota APS (Landau & Ghigo, 1983) are now transforming the situation. These machines are able to measure the positions, magnitudes and image profiles for all objects on a direct Schmidt plate in a matter of hours and for the first time a rapid classification of all the objects above the plate limits in several colours is a practical possibility. Furthermore, the associated computer facilities make possible a rapid integration and intercomparison of the different types of measurements. For instance, objects selected by colour can be further classified by plotting out their prism spectra while in radio source identification programmes, the objective prism spectra and information on the variability of the identifications can be immediately available.

The increased capabilities of Schmidt telescopes have been paralleled by a corresponding improvement in the spectroscopic facilities of large reflectors. New techniques such as image tubes, solid state detectors, fibre optics and on-line data processing have all reduced the time required to obtain spectra for selected objects. Nevertheless, taking even low-resolution spectra of all the QSOs found on even a single deep UKST objective prism plate is still a major undertaking. Taking spectra for all the colour selected QSO candidates is even more formidable but with objective prism data available is no longer necessary. Experiments being carried out with fibre optics on the UKST may further reduce the necessity for follow up spectroscopy on large reflectors (Watson, 1986). To exploit fully the ability of the Schmidt telescope to cover very large areas of sky, investigations must be designed which require a minimum of big telescope time to reach significant conclusions, e.g. the study of QSO clustering and redshift distributions can be confined to emission line QSOs whose redshifts can be obtained directly from the prism plates. Selection procedures must be developed to isolate from the run of the mill QSOs those particular objects which can make the best use of the limited time available on large reflectors. The remarkable amount of detail which can be extracted from prism spectra and which makes such selection procedures possible is probably still not widely appreciated (see Hazard *et al.*, 1986*b*; 1987).

The proper exploitation of Schmidt telescope data could revolutionize many areas of astronomy and lead to exciting new results and discoveries in the years ahead. The progress in the search for very high redshift QSOs using UKST direct and objective prism plates and the facilities of the APM probably provide the best example to date of the capabilites of the new techniques and data processing facilities. The history of the search for high redshift QSOs also illustrates how the earlier approach to QSO surveys, with different groups using a particular technique and each not fully appreciating the results or limitations of the work of the others, has led to much confusion.

2 QSO selection criteria

The story of the search for very high redshift QSOs is a complicated one. To understand it requires some historical background on QSO surveys in general. This, in turn, requires a basic knowledge of the limitations of the early UBV surveys and the clearing up of some misconceptions regarding selection effects in radio and objective prism QSO searches. Part of the complication is caused by the wide variety of QSO spectra which range from strong emission line objects to extreme BAL QSOs whose spectra are dominated by broad absorption troughs and often have only weak emission lines. However, for the purpose of discussing the basic ideas it is sufficient to consider only normal emission line QSOs.

Figure 14.1 is an attempt to show on a single diagram the main features of a typical emission line QSO which are relevant to survey programmes and their relationship to (1) various standard colour filters and (2) the pass bands of the emulsions normally used in objective prism surveys. Because we are mainly interested in high redshift QSOs the spectrum has been drawn for a redshift $z = 2$; slanting broken lines show how the significant features move relative to the filters with changing redshift. The spectrum may be divided into three main sections (1) the region to the red of CIII] where the emission lines are weak relative to the continuum, except for Hα which is visible only at very low redshifts (2) the region from (and including) CIII] to Lyα where strong emission lines dominate the spectrum and (3) the region to the blue of

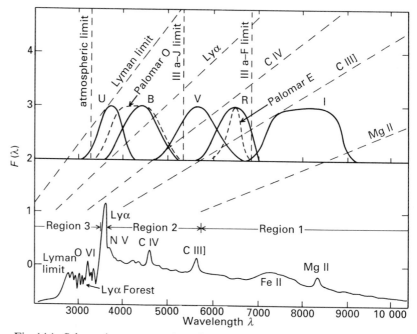

Fig. 14.1. Schematic representation of an idealized QSO spectrum ($F(\lambda)$ vs λ) showing the relationship of its main features to various filters and to the limits of the IIIa-J and IIIa-F emulsions used in objective prism surveys. The spectrum is drawn for redshift $z = 2$. In spectral region 1 the QSO colour is determined mainly by the continuum slope. The blue continuum continues across region 2 but the emission lines of CIII], CIV and particularly Lyα are also important. Region 3 is dominated by Lyα forest absorption and a cut-off at the Lyman limit. The broken lines show how the main features move relative to the filters with changing redshift. The filter responses shown by the solid lines are for the normal UBV system and for the R and I filters commonly used on the UKST; B may be considered equivalent to the J-system on the UKST. The broken lines are for the passbands of the blue (O) and red (E) filters of the Palomar Observatory Sky Survey.

Lyα where there is absorption in the continuum produced by intervening absorption clouds (the Lyα forest lines and damped Lyα systems) and in some, but not all, QSOs a sharp cut-off in the spectrum at or near the Lyman limit. The selection criteria change with redshift as each of these regions occupy the visible part of the spectrum. The characteristics of these regions in turn determine which of the various selection options are most appropriate or convenient at different redshifts. In this section I describe how they apply to (1) the UBV selection technique (2) to radio identifications based on colour selection from the POSS prints and (3) the selection of QSOs from UKST IIIa-J objective prism plates.

2.1 The UBV selection procedure

It is convenient to consider this technique in three stages depending on which of the three spectral regions of Fig. 14.1 is covered by the U, B, and V filters.

(a) z < 0.8

Figure 14.1 shows that for redshifts $z < 0.8$, the U, B and V filters all lie within spectral region (1) to the red of CIII] where the continuum dominates. Except for QSOs with very strong FeII emission the continuum may be considered to have a power law distribution $F(\lambda) \propto \lambda^{\alpha}$ (or $F(v) \propto v^{\alpha-2}$) where α typically lies between 0 and 2.0. As the shape of a power law continuum is independent of redshift then so are the U–B and B–V colours. A flat or rising power law continuum contrasts with a typical stellar spectrum at high galactic latitudes where the Balmer jump and an accumulation of metal lines each result in a depression of the spectral intensity in the UV. This is the origin of the blue U–B colours of QSOs relative to stars with the same values of B–V, i.e. the well known QSO UV excess which is the basis of the UBV colour selection criteria. The U, B and V filters are strategically placed to exploit the change in the slope of the stellar spectra to the blue of $\lambda \approx 4000$ Å. At high galactic latitudes where we are dealing only with old stars of population II, it reduces in practice to a single colour (U–B) selection aimed at selecting objects which are unusually bright in the UV, irrespective of the values of B–V.

The calculated values of U–B for a power law continuum range between U–B $= -1$ for $\alpha = 2$ and U–B $= -0.5$ for $\alpha = 0$ with a median value of U–B ~ -0.8 (Hayman *et al.*, 1979). Very few high latitude stars have U–B < -0.5 which, therefore, forms a useful criterion for the reliable selection of QSOs. There are no serious effects on the colours to be expected from MgII or weaker emission lines but a problem can arise for QSOs with strong FeII emission. A strong FeII emission bump steepens the continuum to the red of MgII and flattens it in the region between MgII and CIII]. This increases the expected UV excess at $z \sim 0.3$ and decreases it at $z \sim 0.6$ and in early optically selected samples of QSOs a peak in the redshift distribution was indeed observed at $z \sim 0.25$ with a broad minimum centred on $z \sim 0.6$.

(b) $0.8 < z < 2.2$

As the redshift increases above $z = 0.8$, region 2 of the spectrum, between CIII] and Lyα, moves progressively across the U,B and V filters. The rise in the continuum level discussed in the previous section continues across region 2 so that the same argument about the continuum colours applies. However, now the effect of the strong emission lines, which are the dominant features of region 2, also have to be taken into account (Strittmatter & Burbidge, 1967). As the lines move through the various filters they perturb both the brightness and the colours of the images; their effect is most significant in the relatively narrow U-filter. The effect of a line raising the brightness in a particular filter is to bring above the plate limit, at that redshift, more QSOs than would otherwise be visible. This has obvious implications as to how to chose the initial list of stellar objects for any colour based survey, but, here, we are more interested in how the lines effect the colour measurements.

If W_R is the rest frame equivalent width of a QSO emission line, the equivalent width (W_z) at redshift z is given by

$$W_z = (1 + z) W_R,$$

and the magnitude change Δm produced by the line in a filter whose effective passband is $\Delta\lambda$ is given by

$$\Delta m \approx 2.5 \log_{10}(1 + (1 + z) W_R/\Delta\lambda),$$

where $\Delta\lambda$ is assumed to be greater than the width of the emission line.

The equivalent width of any particular emission line varies widely from QSO to QSO and there is also a wide variation in the ratios of the lines from different ions. Fairly average values of W_R for Lyα, CIV and CIII] are 75, 25 and 15 Å respectively, although values for Lyα ranging up to 200 Å are not uncommon. Table 14.1 shows the redshifts at which each of these three lines are centred in the U, B and other commonly used passbands and the resulting changes in apparent magnitudes.

The CIII] line has a significant effect on the U−B colour at $z \sim 0.9$ only for very strong emission line objects and even the stronger CIV line has relatively little impact for average line strengths as the brightening it produces in the V-magnitude at $z \sim 1.4$ is partially cancelled by the corresponding brightening in B produced by CIII]. The most significant change is at $z \sim 2$ when Lyα lies in the U-filter and the calculated colour change for the average QSO is $\Delta(U−B) = -0.4$ even after taking account of CIV in the blue filter, for $W_R(Ly\alpha) = 200$, $\Delta(U−B) = -0.8$. The net effect of

Table 14.1. *Magnitude changes (Δm) produced by the CIII], CIV and Lyα lines in the standard UBV filters, the R-filter normally used in the UKST and the O and E filters of the POSS. W_R is the assumed rest frame equivalent width of the line and z the redshift at which it lies near the peak response of the relevant filter.*

| | | U $\lambda = 3690$ $\Delta\lambda = 520$ | | B $\lambda = 4400$ $\Delta\lambda = 990$ | | V $\lambda = 5800$ $\Delta\lambda = 1300$ | | R $\lambda = 6400$ $\Delta\lambda = 900$ | | O $\lambda = 4100$ $\Delta\lambda = 1300$ | | E $\lambda = 6400$ $\Delta\lambda = 450$ | |
		z	Δm	z	Δm	z	Δm	z	Δm	z	Δm	z	Δm
CIII]	$W_R = 15$	0.9	0.06	1.3	0.04	2.0	0.04	2.4	0.06	1.1	0.03	2.4	0.12
	$W_R = 40$		0.15		0.10		0.10		0.15		0.07		0.29
CIV	$W_R = 25$	1.4	0.12	1.8	0.07	2.7	0.07	3.2	0.11	1.6	0.05	3.2	0.23
	$W_R = 106$		0.43		0.28		0.29		0.44		0.21		0.75
Lyα	$W_R = 75$	2.0	0.39	2.6	0.26	3.8	0.27	4.3	0.40	2.4	0.19	4.3	0.69
	$W_R = 200$		0.83		0.59		0.60		0.84		0.46		1.31

the colour changes and the corresponding brightness changes (which do not cancel in the two filters) is a redshift dependent selection which favours the selection of strong emission line QSOs at $z \sim 0.9$ and $z \sim 1.4$ but more importantly favours the selection of even relatively weak line objects at $z \sim 2$. The strong bias at $z \sim 2$ is the reason for the well known peak in the redshift distribution of all early optically selected samples of QSOs and it provides the key to an efficient method of selecting QSOs in any redshift range.

(c) $z > 2.2$

At redshifts above $z = 2.2$, the Lyα emission line moves out of the U-filter into the absorbed continuum of the Lyα forest in region 3 of the QSO spectrum. The fall in continuum level going from red to blue across Lyα is another feature which varies from QSO to QSO. At $z = 2$ it probably averages around 0.1 magnitude but increases with increasing redshift approximately as $(1 + z)^{2.5}$. As the absorbed continuum moves into the blue filter the Lyα emission moves into the B-band where, from Table 14.1, it can produce a brightening (Δm) in B, for an average QSO, of about 0.3 at $z \sim 2.5$; for a QSO with $W_R(\text{Ly}\alpha) = 200$ Å, $\Delta m \sim 0.6$. The combined result of these two effects is to reduce the UV excess and the effectiveness of the UBV selection criteria above $z \sim 2.2$ with the strongest bias now being against rather than for the detection of strong emission line QSOs. However, taking into account the variation in the emission line strengths and continuum slopes among QSOs the bias against the detection of QSOs above $z \sim 2.2$ is not so strong as is generally supposed at least up to $z \sim 2.8$ when Lyα is near the peak response of the B-band and the Lyman limit has progressed significantly into the U-band. By $z \sim 3$, Lyα is beginning to pass out of the B-band but the heavy absorption at the Lyman limit has reached the peak of the U response and not only is the U–B colour excess criterion then no longer valid but the QSO becomes faint in both U and B. Because the main problem for $2.2 < z < 3$ is Lyα in the B-filter, there are obvious advantages in this redshift interval in basing the selection on U–V or U–R rather than U–B but at higher redshifts the simple one colour criterion breaks down and two colour surveys using various combinations of U, B, V, R and I are required.

(d) The final QSO sample

A UBV selected sample of QSO candidates is both incomplete and contaminated by galactic stars, the contamination occurring because of a real overlap in the U–B colours of QSOs and stars and also because of errors in the photometry. The QSOs have to be separated from the stars by follow-up spectroscopy. If, as in the early surveys, the spectroscopy has to be carried out using large reflectors then the limited time available sets a limit to the size and completeness of the final QSO list. Relatively uncontaminated initial samples can be obtained by adopting a very strict colour criterion, say U–B < − 0.7, but only at the expense of excluding a large fraction of the QSOs in the survey area. A less restrictive criterion, say U–B < − 0.3, includes more QSOs but relatively more contaminating stars. In any given survey a compromise, which will depend on the photometric accuracy, has to be struck between the two conflicting requirements.

The limitations imposed by the need for follow-up spectroscopy have eased in recent years by the improved instrumentation at large telescopes and the availability of objective prism material, but particularly by the introduction of fibre optics. With systems now available which can take up to 50 spectra simultaneously, contamination by stars is no longer such a serious problem. However, while work is proceeding to apply the technique to the UKST (Watson, 1986) its present use on large reflectors restricts its usefulness to deep surveys over small regions of sky and it is superior to single object spectroscopy only if the surface density of the objects of interest is such that at least several candidates fall within the limited field of view of the reflector. It is valuable as an adjunct to the UBV surveys because the surface density of faint QSOs with $z < 2.2$ is indeed high. Recent observations show that the surface density of these UV excess QSOs rises from around $1 \deg^{-2}$ brighter than magnitude $+ 18$ to about $40 \deg^{-2}$ at magnitude $+ 21$, while galactic stars with similar colours at these magnitudes are relatively rare. It is this high surface density, relative both to the size of a Schmidt plate and to the density of possible contaminating stars, which is the real key, not just to the use of fibre optics but to the success of the UBV search technique itself. If there were only one or two UV excess QSOs per Schmidt plate it is very unlikely that the UBV technique would ever have been developed.

2.2 Early radio source identification programmes

UBV colours were the basis of some of the early searches for radio QSOs but the more usual criterion was based on the observation that the first few identified QSOs all appeared blue on the POSS, that is their images on the blue O-plates appeared brighter than their images on the red E-plates. This is a less rigid restriction than the UV excess criterion but gave reliable identifications with QSOs even when the radio error boxes were as large as 1 arcmin² or more. The statistical basis for the success of the method was that about 20% of bright radio sources are associated with QSOs above the limits of the POSS while the probability of a blue star falling by chance in an area of 1 arcmin² is only $\sim 10^{-2}$. A sample of 500 radio sources would, therefore, yield about 100 suggested QSO identifications with blue stellar objects (BSOs) with no more than about 5 being contaminating blue stars. Reliable identifications with red stellar objects, on the other hand, require positional accuracies of the order of arc seconds, and as I have already pointed out, reliable identifications were particularly important when spectroscopy still relied upon the photograhic plate.

Selection by colour on the POSS is subject to the same basic redshift dependent colour selection effects as the UBV technique but with the significant difference that emission lines in the broad O-plate response are less troublesome than lines in the narrower U-filter and only the effect of the strong Lyα line need be considered. The O-plate response is centred about $\lambda = 4100$ Å and has an effective bandwidth $\Delta m \sim 1300$ Å. An average Lyα line with $W_R = 75$ Å would therefore produce a maximum increase in brightness of only 0.2 mag and even for $W_R = 200$ Å no more than 0.4 mag. This enhancement by Lyα has to be balanced against the reduction in brightness produced by the absorption in the Lyα forest which lies in the wide O-filter at the same time as the Lyα emission. The net result in that for a QSO with average emission lines, there is little or no effect on the O-magnitude by either Lyα emission or Lyα forest absorption until Lyα passes out of the O-response at about 4900 Å and there is no serious redshift dependent colour selection at least up to $z \sim 3$ (Hayman, Hazard & Sanitt, 1979). Any bias will be in favour of selecting QSOs with $2 < z < 3$ and will be stronger the larger the equivalent width of Lyα. This contrasts with the UBV technique where the bias is against QSOs with $z > 2.2$ and against strong emission line objects. This difference between the use of UBV colours and the method actually used by radio astronomers has not been sufficiently appreciated.

For increasing redshifts above $z = 3$, the O and E colours will become progressively redder. The colour change will be greatest for QSOs with a cut-off near the Lyman limit which should lose their blue colour by the time the Lyman limit has progressed across the O-response to $\lambda \sim 4100$ Å or $z = 3.5$. Significant colour changes above $z = 3$ will also arise from the passage of the

CIV and Lyα lines through the relatively narrow ($\Delta\lambda = 450\,\text{Å}$) E-plate response. The estimates for Δm in Fig. 14.1 show that the CIV line in an average QSO will produce a brightening in the E-magnitude of $\sim 0.2\,\text{mag}$ and the average Lyα line a brightening of $0.7\,\text{mag}$ but that for strong emission line QSOs the corresponding changes could be as large as $0.8\,\text{mag}$ and $1.3\,\text{mag}$ respectively. Strong emission line QSOs should, therefore, appear red around $z \sim 3.2$ and could be very red indeed at $z \sim 4.3$ when the colour change will be further enhanced by the increasing Lyα forest absorption with increasing redshift.

2.3 Objective prism surveys

All the major Schmidt telescope objective prism surveys have used either IIIa–J or IIIa–F photographic plates which can give prism spectra over the wavelength intervals 3400–$5300\,\text{Å}$ and 3400–$6900\,\text{Å}$ respectively, i.e. from just above the atmospheric cut-off to the red cut-offs of the respective emulsions. However, filters are often used to limit the spectra in the blue so as to reduce the sky background and hence improve the limiting magnitude. The grism surveys have used both photographic plates (Hoag and Smith, 1977; Osmer, 1982) and a CCD detector (Schmidt, Schneider & Gunn, 1986). A grism on a large reflector can reach a somewhat fainter limiting magnitude than an objective prism–Schmidt telescope combination but in searches for rare objects any advantage this confers is far outweighed by the much larger field of the Schmidt telescope.

The unfiltered IIIa–F surveys of Hoag & Smith (1977) and Osmer (1982) on the CTIO 4-m telescope reached to magnitude $+21$ with a 30 min exposure, but over an area per exposure of only $0.25\,\text{deg}^2$. The CCD survey of Schmidt, Schneider & Gunn (1986) on the Palomar 5-m telescope reaches to the fainter limit of magnitude $+22$ in a 20 min exposure over the spectral range 4800–$7100\,\text{Å}$ but only over an area per exposure of $8 \times 10^{-3}\,\text{deg}^2$. On the other hand the UKST which covers $36\,\text{deg}^2$ per exposure reaches to magnitude $+20$ in 60 min using unfiltered IIIa–J plates and to $+19$ in 15 min using unfiltered IIIa–F plates; using filtered IIIa–F plates with a blue cut-off at $4900\,\text{Å}$ it reaches magnitude $+20$ in 60 min and covers essentially the same wavelength range as the Palomar Survey. It would take 4500 Palomar exposures to cover the area of a single Schmidt plate and require 1500 hours of 5-m telescope time, at least all the dark time for an entire year. The argument is that going deeper compensates for the decreased sky

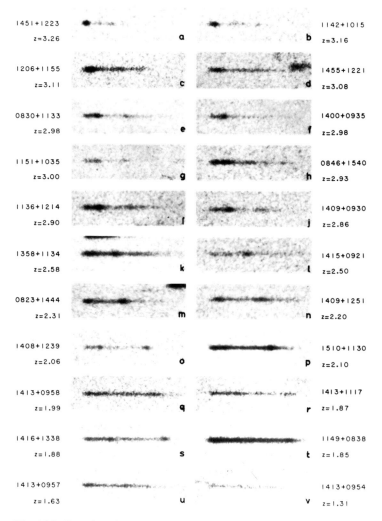

1451+1223
z=3.26   a

1206+1155
z=3.11   c

0830+1133
z=2.98   e

1151+1035
z=3.00   g

1136+1214
z=2.90   i

1358+1134
z=2.58   k

0823+1444
z=2.31   m

1408+1239
z=2.06   o

1413+0958
z=1.99   q

1416+1338
z=1.88   s

1413+0957
z=1.63   u

b   1142+1015
     z=3.16

d   1455+1221
     z=3.08

f   1400+0935
     z=2.98

h   0846+1540
     z=2.93

j   1409+0930
     z=2.86

l   1415+0921
     z=2.50

n   1409+1251
     z=2.20

p   1510+1130
     z=2.10

r   1413+1117
     z=1.87

t   1149+0838
     z=1.85

v   1413+0954
     z=1.31

Fig. 14.2. Samples of IIIa-J objective prism spectra showing how the appearance of the spectra vary with redshift. Note how Lyα becomes easier to detect to the higher redshifts. Wavelength increases to the left. The right hand edge of the diagrams corresponds to $\lambda \sim 3300$ and the red limits of the spectra are at $\lambda \sim 5300$ Å. (Reproduced by courtesy of the UK Schmidt Telescope Unit and the Monthly Notices of the Royal Astronomical Society.)

coverage by virtue of a higher density of faint QSOs. However, even assuming a factor of 5 increase in the number of QSOs per magnitude interval the Schmidt telescope, for the detection of the same number of QSOs, still wins out by a factor of ~ 50 over the Palomar CCD survey and by a factor of ~ 10 over the CTIO Survey; for a flat luminosity function the corresponding factors are ~ 1000 and ~ 50, respectively.

Objective prism surveys have a limited application at low redshifts but become competitive with UBV surveys when spectral region 2 of Fig. 14.1 enters the passbands of the IIIa–J and IIIa–F emulsions. The QSOs become increasingly easy to recognize as CIII], CIV and particularly Lyα successively enter and then move across the prism spectra at redshifts of $z = 0.8$, $z = 1.2$ and $z = 1.8$ respectively. Fig.14.2 shows a sample of UKST IIIa–J unfiltered objective prism spectra which illustrate the changing appearance of the spectra with increasing redshift from $z = 1.3$ to $z = 3.3$ when Lyα passes beyond the red limit of the response of the IIIa–J emulsion. The spectra which were taken using the low dispersion, 2400 Å mm^{-1}, 44 arcmin prism (Nandy *et al.*, 1977) are taken from Hazard *et al.* (1986*b*). These spectra show that at least for low dispersion UKST prism spectra there is not, as appears to be widely believed, an increasing bias against the detection of QSOs with increasing redshift above $z \sim 2$. At least for UKST material, emission line QSOs become more prominent above $z \sim 2.5$ and remain obvious up to $z = 3.3$, the limit set by the red cut-off in the response of the IIIa–J emulsion. Indeed QSOs with $z \sim 3$ are among the easiest of all spectra to recognize on these plates. The increasing visibility of the emission lines with increasing redshift is partly due to the $(1 + z)$ increase in the equivalent width and partly to the fact that the broad QSO lines only become well matched to the resolution of the spectra towards the red limit of the IIIa–J emulsion (Hazard *et al.*, 1986*b*).

The spectra in Fig. 14.2 show that as Lyα moves to the red, the appearance of the spectra becomes increasingly dominated by a drop in continuum level produced by absorption in the Lyα forest. This drop in continuum level both enhances the visibility of the Lyα emission and, when only one line is visible on a spectrum, it provides the main evidence that it is indeed Lyα. Supporting evidence may be provided in some cases by OVI emission, strong absorption features in the blue continuum and as in spectrum 2(*a*), a sharp cut-off in the continuum near the expected position of the Lyman limit. The absorption to the blue of Lyα which is deleterious to UBV photometry therefore becomes an asset in objective prism surveys.

The spectra of high redshift QSOs found on the UKST IIIa–J prism plates provided the experience in analyzing prism plates for the recent successful searches for very high redshift QSOs using UKST IIIa–F plates which in principle extend the redshift range for the detection for Lyα up to $z \sim 4.6$.

3 The search for very high redshift QSOs $z > 3.5$

3.1 Some history

High redshift QSOs are important because of the information they can provide about the structure, early history and evolution of the Universe and the bright ones are particularly valuable as probes of the intergalactic medium. Therefore, immediately following the identification of 3C273 in 1963, the search was on for QSOs at higher and higher redshifts. At first progress was rapid and within a year or so, redshifts up to $z \sim 2.2$ were relatively common. In succeeding years the total of known QSOs with $z < 2.2$ continued to grow but progress to higher redshifts was tantalizingly slow and after 10 years the redshift limit had reached only to $z = 2.88$ (Lynds & Wills, 1970) and almost all those with $z < 2.2$ had been found in radio source surveys. However, 1973 saw a major breakthrough with the measurement in quick succession of redshifts greater than three for two radio QSOs, $0642 + 449$ (OH471) with $z = 3.40$ and $1442 + 101$ (OQ172), with $z = 3.53$. The letters describing these measurements (Carswell & Strittmatter, 1973; Wampler *et al.*, 1973) give little of the background as to how these objects were selected nor do they make clear that they were selected specifically as candidate QSOs with $z > 3$.

It had been recognized as early as 1967 (Bahcall & Sargent, 1967) that a UV excess was not a necessary requisite of a QSO, but as of 1973 essentially all suggested QSO identifications from the POSS were still with blue stellar objects. The association between QSO and BSO had become so close that the terms had become practically synonymous and although some identifications had been suggested with red stellar objects their possible implication does not seem to have been recognized (Gearhart *et al.*, 1972). Also most workers, myself included, were unwilling to make identifications with neutral or red stellar objects unless the positional agreement could be shown to be within the order of 1 arcsec. However, when I heard in 1972 that large numbers of radio identifications based on very accurate radio positions had been made by Dr Browne at Jodrell Bank and by Dr Gent at RRE, Malvern it seemed that the time was right to drop colour selection and, on the basis of the arguments outlined earlier, to try for a significant jump above the existing redshift limit. I contacted Drs Browne and Gent who kindly sent me enlargements from the POSS O and E prints of several of their identifications from which $0642 + 449$ and $1442 + 101$ were selected as particularly interesting candidates which appeared to have either neutral or red colours.

(The image of 1442 + 101 looked brighter on the O print but was classified as neutral from a comparison with surrounding stars all of which also appeared as blue.) The identification of 0642 + 449 from Dr Browne was sent to Dr Strittmatter at Steward Observatory and the identification of 1442 + 101 from Dr Gent to Dr Wampler at Lick Observatory. Within a few weeks a letter from Dr Strittmatter informed me that 0642 + 449 had $z = 3.40$ and shortly afterwards a call from Dr Wampler that 1442 + 101 had $z = 3.53$. I have given these details because the contributions of Browne and Gent have not been adequately acknowledged. The details behind the colour selection were, however, outlined by me in an initialled article in *Nature* accompanying the Carswell and Strittmatter letters (Hazard, 1973; see also Gent *et al.*, 1973). The situation was confused because the identification for 0642 + 449 had already been proposed independently by Gearhart *et al.* (1972) and that of 1442 + 101 by Véron (1971). Neither paper suggested that the objects might be high redshift objects nor could the identifications be considered as reliable as those based on the very accurate positions of Browne and Gent.

That the first two objects selected from their neutral or red colours should be high redshift QSOs led to optimism that the discovery of even higher redshifts would soon follow. However, these high hopes failed to materialize and it was to be several years before we again exceeded even $z = 3$. This despite the fact that the new aperture synthesis instruments at Cambridge and Westerbork and interferometers such as the three element instrument at Greenbank were making it much easier to obtain accurate radio positions. I had embarked at that time with H.S. Murdoch on a programme aimed at obtaining identifications for sources in the MC2 and MC3 catalogues with spectroscopy being carried out at Lick by E.M. Burbidge, J. Wampler, J. Baldwin and E.H. Smith. We made a particular effort to remove colour selection and noted all stellar objects near the radio positions irrespective of colour. For objects which could not be identified from the catalogue positions we measured more accurate positions to about ± 1 arcsec either at Westerbork (with R. Ekers) or at Green Bank (with J.J. Condon). However, all the red or neutral stellar objects we examined turned out to be galactic stars or BL Lac objects. It was not until we directed our attention to an extension of the MC2 and MC3 catalogues (commonly called MC5) in a small 1.5 deg strip between RA 8^h and RA 10^h that we had any success. The distribution of redshifts in this region turned out to be very different from those in the MC2 and MC3 catalogues. In the MC2 and MC3, one third of all QSOs had $z < 1.5$. In MC5, two-thirds had $z > 1.5$ and of the first few

MC5 sources we examined one, $0938 + 119$, had $z = 3.19$ (Beaver *et al.*, 1976), a second $z = 2.77$ and a third $z = 2.99$ (Baldwin *et al.*, 1976). The explanation for this difference could not be sought in terms of the lower flux limit of MC5 because all were above the limit $S = 0.45$ Jy (at 408 MHz) of MC2 and MC3 and the observations clearly raise the question of clustering of high redshift QSOs. (Hazard, 1977*a*).

The new high redshift objects were all blue on the POSS except for $0938 + 119$ which was distinctly red; however, its unusual colour, much redder than $1442 + 101$ at $z = 3.53$, was easily explained by its unusually steep continuum and by CIV lying in the relatively narrow E plate response. Otherwise, the colours confirmed the conclusion reached earlier that selecting on the basis of a blue colour on the POSS does not discriminate against objects up to $z = 3$. It furthermore demonstrated that the absence of objects above $z = 2.2$ in the large ($\sim 500 \deg^2$) survey area of MC2 and MC3 could neither be attributed to colour selection nor to any deficiency in our identification procedure. The steep fall from $z \sim 2$ to $z \sim 2.2$ rather suggested a real switch on time for QSOs or at least an epoch of maximum QSO formation (Hazard, 1977*a*). The fall can be explained in terms of the optical and radio limits of the sample combined with a steep QSO luminosity function without any need to invoke a radical change in the QSO distribution at this redshift but not if the rising QSO density with increasing redshift inferred at lower redshifts (Schmidt, 1968) extends past $z = 2$ (Hazard, 1977*b*) which still implies that $z \sim 2$ is a significant one in QSO evolution. I do not want to go into this aspect here but I will return to the observations later because they determined my approach to optical surveys for very high redshift objects. For now, I merely point out that the early discovery of two QSOs with $z \sim 3.5$ had the unfortunate consequence of focussing attention on redshifts $z > 3.5$ when it would have been more profitable, for a time at least, to concentrate on redshifts around $z \sim 3$.

With the development of the new aperture synthesis instruments and new interferometers, the measurement of radio positions accurate to ± 1 arcsec became routine and colour selection was soon eliminated completely from radio source identification programmes. However, the introduction of slitless spectroscopy with its bias to redshifts > 2 appeared to be an even more promising approach. With its introduction, astronomers again anticipated the discovery of very high redshift QSOs in large numbers, particularly with the early success of the technique in the discovery of the $z = 3.45$ QSO 2227-3928 (Hoag & Smith, 1977; Smith *et al.*, 1977) which was found in a survey covering 5.1 $\deg^2$ of sky using a IIIa–F emulsion and a

Table 14.2. *A list of QSOs with z ⩾ 3.5 as of September 1986. For optically selected QSOs the source of the plate material is shown and, where appropriate, the measuring machine used*

Name	z	$m(R)$	Survey tech	Confirmation	Ref.
2239-3608	3.50		Prism UKST	Sept. 1986 CTIO	1
0105-2634	3.50	17.5	Prism UKST	1985 AAT	2
1159-1223	3.51	17.5	Prism UKST	1981 MMT	3
OQ172	3.53	17.0	Radio	1973 Lick	4
PKS????	3.55	19.0	Radio	Aug. 1986 AAT	5
Q1409 + 732	3.56	19.0	Grism KPNO	1986 KPNO	6
DHM0054-284	3.61	18.0	Colour UKST + COSMOS	1983 AAT	7
0055-2659	3.67	17.0	Prism UKST	1984 INT	8
PKS1351-018	3.71	19.5	Radio	1986 AAT	9
PKS2000-330	3.78	17.0	Radio	1982 AAT	10
PKS????	3.79	17.0	Radio	Aug. 1986 AAT	5
1208 + 1011	3.80	17.5	Prism UKST	Jan. 1986 Palomar	11
0135-4239	3.97	19.0	Colour UKST + APM	Sept. 1986 CTIO	12
0046-2919	4.01	19.0	Colour UKST + APM	Aug. 1986 AAT	13

(1) Hazard and McMahon (1987); (2) Hazard, McMahon & Morton (1987); (3) Hazard *et al.* (1984); (4) Wampler *et al.* (1973); (5) Jauncey *et al.* (1987), private communication; (6) Anderson & Margon (1987); (7) Shanks, Fong & Boyle (1983); (8) Hazard & McMahon (1985); (9) Dunlop *et al.* (1986); (10) Peterson *et al.* (1983); (11) Hazard, McMahon and Sargent, (1986a); (12) McMahon *et al.* (1988); (13) Warren *et al.*, (1987).

grism on the CTIO 4-m telescope. However, once again the early promise was not fulfiled and by 1984, apart from a $z = 3.5$ QSO discovered in 1981 on a UKST IIIa–F low dispersion objective prism plate (Hazard *et al.*, 1984), the technique had yielded no higher redshifts. When a second CTIO survey, with an improved prism also failed to turn up any higher redshifts, it was suggested that there is a cut-off in the QSO distribution at $z \sim 3.5$ (Osmer, 1982). However, within a few months of Osmer's paper it was demonstrated that QSOs both exist and are detectable at higher redshifts by another success for the radio selection technique, namely the identification of PKS 2000–330 with a $z = 3.78$ QSO (Peterson *et al.*, 1983) and the discovery of a $z = 3.61$ QSO, DHM 0054–284 where the primary selection was based on a set of U, B, V, and R plates processed on COSMOS (Shanks *et al.*, 1983). Since then the floodgates have opened and QSOs with $z > 3.5$ are being discovered at an increasingly rapid rate. This is illustrated by Table 14.2 which lists all such QSOs of which I am aware, together with their date of discovery and the technique employed.

Table 14.2 comprises a total of 14 QSOs of which five, including four of the highest redshifts known have been discovered in the past eight months. A notable feature is that all but one of the optically selected QSOs have been found using UKST Schmidt telescope material and that most are relatively bright. The success using the Schmidt telescope contrasts with the very limited success of deeper searches over smaller regions of sky using either slitless spectroscopy or colour techniques. Only one QSO at $z = 3.56$ (Anderson & Margon, 1987) has been found using a grism on a large reflector. The IIIa–F CTIO prism survey of Osmer (1982) covered 5 $\deg^2$ of sky and reached to magnitude $+ 21$ but found no examples of Lyα emission in the range $2.7 < z < 4.7$ although QSOs were found by their CIV emission in the interval $2.77 < z < 3.36$. The Palomar survey which reached to magnitude $+ 22$ and covered 1 $\deg^2$ found only QSOs in the range $0.9 < z < 2.66$. Likewise the deep multicolour surveys described by Koo (1983) and Koo *et al.* (1986) found no QSOs with redshifts significantly above $z = 3$ to a limiting magnitude B ~ 23. Calculations based on the luminosity function models of Schmidt and Green (1983) had suggested that significant numbers of high redshift QSOs should have been found in all these deep surveys. As already noted these results had led Osmer (1982) to conclude that there is a cut-off in the QSO distribution at $z \sim 3.5$ while Schmidt *et al.* (1986) have concluded that QSOs with an absolute magnitude $m_B = - 25$ suffer a redshift cut-off near or below a redshift of 3.

3.2 High redshift QSOs as very rare objects

The difficulty in finding QSOs with $z > 3.5$, particularly in searches over small regions of sky, has to be considered in the light of the fact that even QSOs in the redshift range, $3 < z < 3.3$ are not exactly common. Although even weak emission line QSOs in this redshift range are readily found on Schmidt telescope IIIa–J prism plates, and such plates covering several thousand square degrees have now been searched, the latest QSO catalogue (Hewitt & Burbidge, 1987) lists only about 100 in this redshift interval. A detailed search of a UKST plate centred on RA 1500 Dec $+ 1100$ confirmed a total of 9 with B(J) < 20, about a factor of 10 per unit area less than the number found around $z \sim 2$ in the same redshift interval (Hazard, 1985). A similar steep fall has been noticed in all objective prism surveys but has often been represented as at least partly due to a selection bias which increases with increasing redshift. The steep fall at $z \sim 2$ in UBV selected samples is also usually attributed to redshift dependent selection. As I have already argued the selection biasses, particularly for the low dispersion UKST prism surveys, have been over-emphasized. Part of the widespread failure to recognize this must be attributed to the early discovery of the $z = 3.53$ QSO 1442 + 101 focussing too much attention on redshifts $z > 3.5$ when an investigation of the more easily found $z = 3$ QSOs would have shown that the decline in QSO numbers had set in well before $z = 3$ (Hazard, 1985; Hazard & McMahon, 1985). In fact the steep decline was already obvious from the radio surveys of the 1970s. It should be noted that while the low dispersion UKST prism is ideally suited to the detection of QSOs around $z \sim 3$, those surveys using higher dispersion prisms or grisms really may be biassed against the higher z QSOs and that a lack of high z objects in such

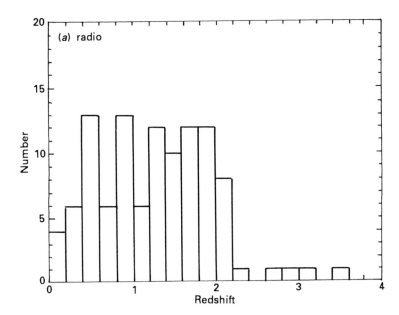

Fig. 14.3. Histograms showing the distribution of QSOs as a function of redshift for (*a*) a sample of Molonglo radio sources (*b*) a IIIa-J objective prism survey over an area of ~ 6 degrees^2 and (*c*) the UBV selected sample of Marshall *et al.* (1983). Note the fall in numbers at $z \sim 2.2$ in all three plots despite the different selection effects involved.

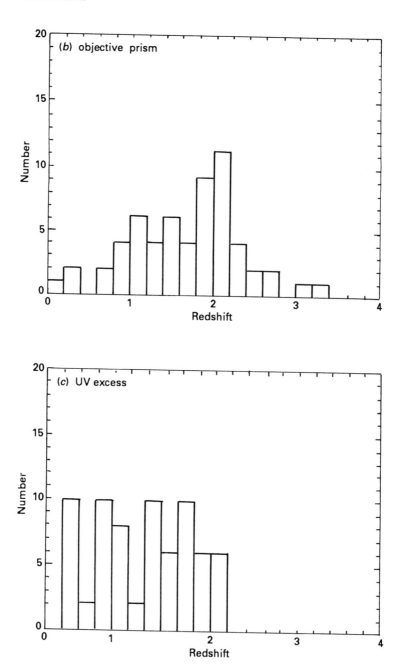

Fig. 14.3 (*Cont.*)

surveys does not necessarily imply a real decline in QSO counts above $z \sim 2.2$.

The reality of the steep fall in QSO numbers is demonstrated quite clearly by the redshift distributions shown in Fig. 14.3. Fig. 14.3(a) is a plot for Molonglo radio QSOs, but updated from the earlier version (Hazard, 1977a) referred to earlier; it is complete to magnitude ~ 19.5. Figure 14.3(b) shows the results of a UKST IIIa-J objective prism survey by myself and W.L.W. Sargent over an area of 6 deg^2. It is a revised version of an earlier plot (Hazard, 1985) and is essentially complete to $m(J) = 19.5$ from $z = 1.8$ to $z = 3.3$. Fig. 14.3(c) is the corresponding distribution for the UBV colour selected sample of Marshall *et al.* (1983) which is complete to B $= +19.8$. All show the same steep fall in the counts at $z \sim 2.2$ despite their very different selection biasses. The colour selected sample is biassed against redshifts with $z > 2.2$, the prism sample favours redshifts $z > 1.8$ while the radio sample is essentially redshift independent. That the cut-off occurs at the same redshift in all samples shows conclusively that it must represent a real fall in the QSO counts rather than any artifact in the selection procedure. Even the UBV selected sample argues for a real decline at $z > 2.2$ for, given the wide variation in continuum slope and emission line strength among QSOs, there is no way that the selection bias could produce such a precipitous fall unless the real counts were also falling steeply.

Given the small numbers of objects with $z \sim 3$, the rate of decline (for a magnitude limited sample) indicated by the prism survey is in good agreement with that indicated by the radio data. The prism survey shows a decline in the counts per unit of redshift by a factor of 10 between $z = 2$ and $z = 3$ while the radio survey shows a decline by a factor of 20. It has already been suggested that the objective prism data indicates a real fall in the co-moving density of QSOs (Hazard, 1985). However, all I wish to do here is to point out that the radio data presented in 1977 had already shown quite clearly that high redshift objects are very rare. Thus, the total Molonglo survey covers a total area of about 500 deg^2 and is certainly essentially complete up to $z \sim 3.5$. Two QSOs were observed with $3 < z < 3.53$ which corresponds to 1 source per 250 deg^2. Taking the radio QSOs with $S > 0.5$ Jy at 408 mHz to form 5% of the total QSO population gives a total QSO density for $3 < z < 3.5$ of about 1 QSO per 12 deg^2 to magnitude $\sim +19.5$. Assuming the decline to continue at the same rate above $z = 3.5$ would indicate probably no more than 1 QSO per 30 deg^2 brighter than 19.5 for $3.5 < z < 4.5$. Even assuming an increase of a factor of 5 per magnitude interval, an optimistic assumption, this corresponds to only about 0.3 QSO deg^{-2} to magnitude $+21$ and 1.5 QSO deg^{-2} to magnitude $+22$. These estimates are in reasonable agreement with later estimates from objective prism surveys of 1 QSO per 6 deg^2 with $m(R) < 19.3$ in the interval $3.3 < z < 3.9$ which for a flat luminosity function extrapolated to 0.1 QSO deg^{-2} with $m(R) < 21$ and $4 < z < 5$ of which about half would have $m(R) < 19$ (Hazard & McMahon, 1985). For $3.5 < z < 4.5$, these objective prism estimates correspond to 0.1 QSO deg^{-2} with $m(R) < 19.5$ and 0.2–0.3 QSO deg^{-2} with $m(R) < 21$.

The estimate of 1 QSO per 10 deg^2 brighter than 19.5 in the interval $3.5 < z < 4.5$ is independent of the cosmological model and depends only on the plausible assumption that the decline beyond $z = 3$ continues at roughly the same rate as between $z = 2$ and $z = 3$. Justification for a continued decline is provided by the results from the UKST IIIa-F prism survey (Hazard & McMahon, 1985) and also by the progress of the radio source identifications. The present radio surveys, based on postional accuracies of the order of 1 arcsec are completely free of colour selection yet the QSOs found at $z \sim 3$ continue to outnumber those found at redshifts between 3.5 and 4.

The main uncertainty is in the estimate for the counts at a fainter magnitude limit. Present indications are that at high redshifts the luminosity function is relatively flat so that the counts are only weakly dependent on magnitude limit. Whether this represents a change in the luminosity function between $z = 2$ and $z = 3.5$ needs further investigation. I have always thought it suspicious that counts of QSOs on different UKST plates by different observers selecting on the basis of the emission lines are so similar, 6–10 deg^{-2}; larger numbers have been obtained but by making extensive use of the colour information in the prism spectra. Given the uncertainty in the magnitude limits of the various surveys, the agreement appears inconsistent with a steep luminosity function at $z \sim 2$, the redshift around which most of the QSOs found are clustered. It may be that the steep QSO luminosity function inferred at low redshifts represents only the decline expected near the upper limit of any luminosity function and may persist over a range of only a couple of magnitudes.

Whatever may be the truth about the magnitude distribution it is clear that the radio data available more than 10 years ago indicated that QSOs above $z > 3.5$ could be very rare objects indeed and that the only hope of finding them in reasonably large numbers was to search very large areas. In any case that was the conclusion I reached which is why I have always based my attempts at finding them on the use of Schmidt telescope plates.

3.3 *Optical search techniques for very high redshift QSOs*

Figure 14.1 shows that three of the main characteristics of QSO spectra are (i) the continuum emission to the red of Lyα which, over most of its range, rises more steeply to the blue than that of the majority of galactic stars, (2) the strong Lyα emission line itself and (3) the absorption to the blue of Lyα including a possible cut-off at the Lyman limit. The last two of these regions move completely into the visible region of the spectrum only at redshifts $z > 3$. A QSO at these high redshifts is therefore far more distinctive than at lower redshift, say around $z = 1$, when the visible region is dominated by the continuum emission. If follows that, in principle, QSOs are easier, rather than harder, to detect at high redshifts and the difficulty in finding them is not because they then have less distinctive spectra but, that being so rare, it is more difficult to select them from a proportionately higher number of contaminating stars. If the numbers of QSOs were constant in redshift we would probably be discussing the nature of the low redshift cut-off.

The QSOs at $z < 2$ are easy to select, both because they are common and because their constantly rising continuum allows them to be distinguished from high latitude stars which characteristically have a UV deficiency below $4000\,\text{Å}$. The radio identifications based on the blue colour of the images on the POSS relied more on the overall slope of the continuum. While in itself this does not clearly distinguish QSOs from blue stars, the radio positions reduced the search area so that the probability of a blue star lying within it was small. In principle, the overall blue continuum could still be used at redshifts $z > 3$ by using two filters well to the red of $5000\,\text{Å}$, say by using a two colour system based on R and I magnitudes. Even with present day radio surveys such a procedure could be useful to isolate prospective high z QSOs from a large list of identifications. However, the obvious way to proceed is to use all the information available, namely, the continuum slope, the Lyα emission line and the absorption to the blue of Lyα. The remainder of this paper outlines the principles behind the IIIa–F objective prism and broad band colour surveys which have been responsible for much of the progress over the past two years and which each emphasize different aspects of the QSO spectra. Finally, I present some recent results using medium band filters which I believe offer many advantages over the other techniques.

3.3(a) *UKST IIIa-F objective prism spectra*

The upper redshift limit for detection of Lyα on a IIIa-J objective prism spectrum is $z = 3.3$, when Lyα passes beyond the red limit of the emulsion at ~ 5200 Å. The IIIa-F emulsion which has a red cut-off at about 6800 Å extends the limiting redshift to $z \sim 4.6$. However, while it has been widely used in the CTIO grism surveys it has not usually been considered suitable for QSO surveys using the low dispersion, 2400 Å mm^{-1}, arcmin, prism on the UKST. The low dispersion of this prism above ~ 5200 Å is not, however, a particularly significant feature at least up to $z \sim 4$. Provided the continuum is not saturated, an objective prism spectrum may be considered as an ideal diffraction limited spectrum convolved with a smoothing function whose size, if the object is stellar, is that of the seeing disk. For the best UKST plates the seeing disk is about 1.5 arcsec and the size of the smoothing function varies from about 25 Å at 3400 Å through 90 Å at the red limit of the IIIa-J emulsion to 200 Å at the red limit (6800 Å) of the IIIa-F. In velocity space, the average width of QSO emission lines are ~ 5000 km s^{-1} which corresponds to line widths ranging from 60 through 90 to 120 Å at the same wavelengths. The lines in a typical QSO are, therefore, over-resolved in the ultraviolet and blue and best matched to the seeing disc at the limit of the IIIa-J emulsion at 5200 Å which for Lyα corresponds to $z = 3.3$. This together with the $(1 + z)$ increase in equivalent width explains the increased visibility of Lyα with increased redshift evident in Fig. 4.2. Contrary to popular opinion, even on the low dispersion UKST plates, Lyα is not particularly obvious for $1.8 < z < 2$ where QSOs are in reality more easily recognized by their long and more or less uniform density continuum spectra than by the Lyα line itself. Above $z = 3.3$, the seeing disc is larger than the QSO line widths and the line visibility is decreased by the increase in the ratio of the continuum flux to line flux per resolution element, although the decrease is partly compensated by the $(1 + z)$ increase in equivalent width. If a line is considered, detectable, when it produces an enhancement of 1 magnitude, over the continuum the minimum detectable rest frame equivalent width (W_R) of Lyα rises from about 30 Å at $z = 3.3$ to about 55 Å at $z = 4.6$ which is still well below the upper values of W_R observed in lower redshift QSOs.

Irrespective of the above arguments, the ease with which relatively large numbers of QSOs with $3 < z < 3.3$ were being found on the IIIa-J plates suggested that significant numbers must exist at higher redshifts and that at least the strong emission line examples could be found on IIIa-F plates. The discovery of the $z = 3.51$ QSO $1159 + 1223$ (Hazard *et al.*, 1984) on the first IIIa-F plate examined showed that this was indeed the case. Further experience with these plates has shown that the main problem is not in detecting Lyα emission but in a possible confusion with cool stars. Certain types of cool stars show features in their prism spectra which can be mistaken for Lyα emission at about 6000 Å. However, it has been found possible to achieve essentially 100% reliability in the selection by restricting the searches to $z < 3.9$ and by using auxilliary IIIa-J prism and direct R and J plates to check spectral features and colours of all IIIa-F selected candidates. Moreover, the redshift can be measured directly from the prism spectrum to an accuracy of about ± 0.05.

The restriction to redshifts $z < 3.9$ is not fundamental to the prism technique. It was adopted in the early phases of the survey because of the difficulty of obtaining time on large telescopes to search for objects which most astronomers (or at least those on the time assignment committees) had come to believe did not exist. The objective was to obtain completely reliable identifications. There is, however, no particular difficulty in detecting Lyα at higher redshifts because emission lines in low redshift QSOs, (e.g. CIV in $z = 3$ QSOs) have already been seen in prism spectra at $\lambda \sim 6300$, which for Lyα corresponds to $z = 4.2$. Contamination by cool stars would be a problem but could be reduced with the help of colour information from the J, R and I survey plates. Redshifts above $z = 4.2$ will probably require the use of the UKST 800 Å mm^{-1} prism with a suitable filter to limit the sky background.

Searches on unfiltered, low dispersion IIIa-F prism plates have now been completed to R ~ 18.5 mag in two fields centred on RA 1204 dec 1129 and RA 0055 dec 2803 respectively. Two QSOs with $z = 3.51$ and 3.80 were found on the 1204 + 1129 plate and three with $z = 3.50, 3.61$ and 3.67 are visible on the 0058–2803 plate. These are approximately the numbers expected from an extrapolation of the QSO counts between $z = 2$ and $z = 3$. Searches are now in progress to $m(R) \sim 20$ mag using 60 min exposure filtered IIIa-F plates where the spectra are limited to the interval 4900–6800 Å. The filtered plate of the 0053–2803 field shows very clearly the three QSOs with $z > 3.5$ already known but no obvious fainter candidates at similar redshifts. It seems already clear that in this area of sky, QSOs with $3.5 < z < 3.9$ and $18.5 < m(R) < 20$ do not occur in significantly larger numbers than QSOs of the same redshift with $m(R) < 18.5$. This argues for a relatively flat luminosity function and supports the suggestion that searches for bright high redshift QSOs over large areas of sky are likely to be more productive than deep searches over small areas (Hazard & McMahon, 1985). It also suggests that the filtered plates may have little to recommend then over unfiltered plates. The unfiltered plates have the advantage of a shorter exposure time and also that low redshift QSOs are more easily recognised on them than on the unfiltered plates.

3.3(b) Broad band colour surveys

(1) *The basic technique and early results*

The objective prism surveys for high redshift QSOs rely on the detection of the Lyα emission line with the absorption to its blue and the blue colour of the continuum to its red being used mainly to provide supporting evidence for the line identification. In broad band colour surveys, on the other hand, the latter two features form the basis of the primary selection criterion although the presence of Lyα in one of the filters can be important to its success. The colour change in a QSO spectrum produced by Lyα forest absorption and particularly by absorption at or below the Lyman limit provides an obvious method of separating QSOs from stars by their broad band colours. Not all QSOs show a clear cut-off at the Lyman limit but in most of them the continuum to the blue of the expected position of this limit is very weak. This weak continuum could represent an intrinsic decrease in level determined by the spectrum of the QSO core or could result from the integrated effect of weak Lyman limit absorptions associated with the numerous Lyα forest clouds. However that may be, the result is that a high redshift QSO will be characterized by two colours, a very red colour measured between Lyα and a point in the QSO spectrum well to its blue and a very blue colour measured from Lyα to a position well to its red, this latter colour corresponding to the continuum slope in spectral regions (1) and (2) of Fig.14.1. It is obvious that to measure the colour change for a particular QSO the technique will require three filters straddling Lyα with the central filter ideally placed on or just to the red of Lyα itself. Fig. 14.1 shows that for $3.3 < z < 4$, Lyα lies in the passband of the V-filter so that a suitable choice of filters would be B, V and R or I, while for $4 < z < 4.6$ where Lyα lies in the R-filter an appropriate choice would be B or V, R and I. In principle the U-filter could be used as the bluest of the three filters but, as will be clear from the following discussion, it would offer no advantages over the B-filter at $z > 3.5$ and any advantages it might offer for $3 < z < 3.5$ will be offset by the effects of stellar absorption below $\sim 4000\,\text{Å}$.

Although high redshift surveys are basically two colour surveys requiring the use of three filters, they differ in important respects from UBV surveys for QSOs with $z < 2.2$. The UBV technique exploits the UV absorption in galactic stars rather than any specific feature in a QSO spectrum so that, over the whole range of its applicability, the redshift of the QSO is a secondary consideration; as already described, up to $z \approx 2.2$ the QSO colours are those of the power law continuum of spectral regions 1 and 2 of Fig. 14. 1 and depend only weakly on redshift. The high redshift surveys, on the other hand, specifically exploit the colour change in QSOs over Lyα and any particular set of filters can be used only over a very limited redshift range. To cover a wide range of redshift above $z \approx 3$ requires the use of several filters which may be combined in different ways to search for QSOs in different redshift intervals. The most significant difference, however, is the direction in which the colour changes operate as may be seen by considering a UBV survey for QSOs with $z \sim < 2.2$ and a BRI survey for QSOs with $z \approx 4$. In both surveys, the two reddest filters measure the QSO continuum slope to the red of Lyα so that in both B–V and R–I the relevant QSO will have colours comparable to those of the bluest galactic stars. In the UBV survey, the UV absorption in galactic stars acts to enhance the relatively blue QSO colours and hence produces a clear separation of stars and QSOs in U–B. In the BRI surveys, however, the QSO absorption acts, not to separate the stars and QSOs in B–R but, rather to drive the B–R colours of QSOs away from those of the relatively rare blue stars towards those of the commoner cooler stars. It follows that there is no clear separation of QSOs and stars in either B–R or R–I. However, while the R–I colours of $z \sim 4$ QSOs are independent of their B–R colours, the corresponding colours of galactic stars are strongly correlated. Redder B–R colours imply lower effective black body temperatures and, hence, also redder R–I colours. Therefore, in a two colour (B–R, R–I) diagram, QSOs with very red B–R colours will tend to be bluer in R–I than galactic stars with similar values of B–R. In principle, therefore, high redshift QSOs with $z \sim 4$ can be found from such a two colour diagram by simply selecting the bluest objects in R–I at any particular value of B–R.

It is clear from the discussion given above that the selection of high redshift QSOs is a more complicated matter than selecting lower redshift QSOs by their UV excess. It is further complicated by the fact that while QSOs with $z < 2.2$ are relatively common, high redshift QSOs, particularly above $z \sim 3.5$, are very rare. Thus, while UBV surveys can be carried out over small areas of sky, the high redshift surveys need to be made over areas at least comparable to that of a whole Schmidt plate. Furthermore, the weaker colour separation of the high redshift QSOs and the relatively much larger fraction of possible contaminating stars impose more severe constraints on the accuracy of the magnitude measurements. Simple procedures, such as three colour images on a single plate are no longer practical; I tried it some 10 years ago in collaboration with W.L.W. Sargent but found confusion with cool stars too much of a problem. The difficulties are such that it is only in the past few years, with the development of the high speed scanning machines and appropriate software, that such surveys have become feasible. The increased spectroscopic capabilities on large telescopes, and particularly the development of red sensitive detectors, are also important factors. By making it easier to check selected candidates they allow a lower degree of reliability in the candidate list than could have been tolerated a few years ago. The demonstration by the IIIa–F prism surveys that QSOs with $z \sim 3.5$ existed in reasonably large numbers has also been important, for time assignment committees are now more ready to grant time for follow up spectroscopy on large telescopes.

The first significant success for the machine based broad band colour technique was the discovery of the $z = 3.61$ QSO, DHM0054–284 (Shanks *et al.*, 1983) in the same SGP field in which the $z = 3.67$ and $z = 3.50$ QSOs 0055–2569 and 0105–2634, were later found in the IIIa–F prism surveys. It was selected from a set of UJVR plates from the UKST which were analyzed on COSMOS, the selection being based on its red colour in J–V and blue colour is V–R. The next important steps were also based on UKST material but with the analysis being carried out on the Cambridge APM machine. Firstly, there was the demonstration by Irwin *et al.*, (1985) that both DHM0054–284 and 0055–2659 could be separated from galactic stars on a (J–V, V–R) two colour diagram. This was soon followed by the discovery, also in the SGP, of the faint (R > 19) QSO 0046–293 with $z = 4.01$ (Warren *et al.*, 1987) in a multi-colour survey based on two plates in each of the five passbands U, J, V, R, and I. The $z = 4.01$ QSO lies well separated from the galactic stars in a region in a two colour (J–V, V–R) diagram where objects around $z = 4$ might be expected to occur (Irwin *et al.*, 1985). In addition the survey yielded a faint (R > 19) $z = 3.42$ QSO 0046–276, and also showed the already known $z = 3.61$ and $z = 3.67$ QSOs well separated from the galactic stars. The use of two plates in each passband is an important feature of this survey, the improved colour estimates reducing the possible contamination by stars with spurious colours. The rationale behind the selection procedure was that QSOs would lie in low density regions of the 4–D space defined by the four colours (U–J, J–V, V–R, and R–I) and consisted of a search for any object with a combination of four colours that is noticeably different from all conmon stars. However, it is not so much the 4–D aspect that is important but that, as discussed above, different combinations of colours are required in different redshift intervals. This has been demonstrated by McMahon *et al.*, (1988) who, within a month of the discovery of 0046–293, found a $z = 3.97$ QSO, 0135–4239, using only a two colour plot based on J, R and I; this work also showed that, at least around $z \sim 4$, the APM photometry is sufficiently accurate to allow the reliable selection of QSOs based on only one plate in each passband. However, the movement of candidate objects relative to the sequence of stars from one colour diagram to another does provide the means of obtaining a more reliable set of selections then can be obtained using a single two colour diagram. However we may choose to look at the analysis, these colour surveys using UKST material, analyzed on the APM, represent a very significant step forward in high redshift QSO surveys.

(2) *Selection effects in broad band colour surveys*

The empirical procedures adopted in the high redshift colour surveys described above have proven highly successful but give no idea of the selection effects involved. Some of the considerations which underline the colour technique were outlined at the beginning of the previous section. In this section the technique is analyzed in more detail using schematic two colour diagrams in (B–V, V–I) and (B–R, R–I). In the schematic (B–R, R–I) diagram shown in Fig. 14.4(a), the hatched area represents the sequence of galactic stars while QSOs are represented by solid circles. Each QSO is considered to have a spectrum similar to that sketched in Fig. 14.1 with a strong cut-off in the continuum at the Lyman limit. However, the emission lines will be ignored except for Lyα which has been assumed to have a rest frame equivalent width $W_R = 75$. Firstly, we note that up to $z = 2.2$, the three filters all lie in spectral regions (1) and (2) of Fig. 14.1 and to the red of the Lyα emission. Therefore, all QSOs with $z < 2.2$ will be blue in both B–R and R–I and lie in the region indicated by the broken circle. The scatter of the points within this broken circle is assumed to be real so that their different positions represent real differences in the colour of their continuum spectra. We now consider how objects at different positions within the circle would move if they were to be shifted to higher redshifts. In particular we consider five QSOs: QSO 'a' shifted to $z \sim 2.5$, 'b' and 'c' each moved to $z \sim 3.6$, 'd' to $z \sim 3.9$ and 'e' to $z \sim 4.3$; the colours of the QSOs 'b' and 'c' correspond to continuum slopes which respectively are redder and bluer than the average.

Figure 14.1 shows that up to $z \sim 3.6$ the R and I filters measure only the continuum slope above Lyα so that the R–I colours are independent of redshift. The colour changes of QSOs 'a' to 'c' will therefore occur only in

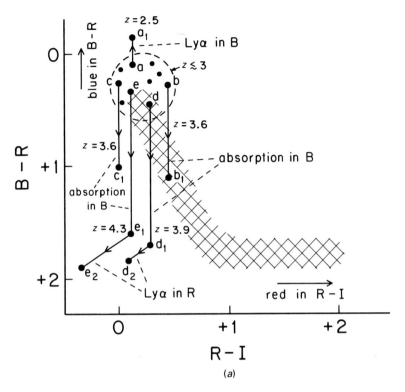

Fig. 14.4. Schematic (B–R, R–I) and (B–V, V–I) diagrams illustrating the principles behind the colour selection of high redshift QSOs. In general QSOs with $z < 3$ lie in the region of the broken circle at the upper left hand corner of both diagrams but Lyα in the B-filter will move QSOs around $z \sim 2.5$ bluer in both B–V and B–R. Objects marked 'a' to 'e' represent QSOs with $z < 2$. The solid lines show how these QSOs would move in each of the two colour diagrams if they were shifted to redshifts 2.5, 3.6, 3.9 and 4.3 respectively. The displacement 'a' to 'a_1', shows the effect of Lyα in the B-filter; otherwise, the vertical displacements to positions subscripted (1) arise from absorption, mainly near the Lyman limit, in the B-filter. In Fig. 14.4(a), displacements from positions subscripted (1) to subscript (2) arise from Lyα in the R-filter, while in Fig. 14.4(b) they arise from Lyα in the V-filter. The displacements shown are for a rest frame equivalent width for Lyα of 75 Å. The hatched area represents the region occupied by the main sequence stars.

B–R and result from the movement of first Lyα and then the absorption features of spectral region (3) through the B passband. The way in which the colour changes operate are indicated by solid lines in Fig. 14.4(a). For red-shifts $2.2 < z < 3$ the most significant feature is the passage of Lyα through the B-filter which produces bluer colours in B–R. The contribution from Lyα reaches a maximum of about 0.3 mag (Table 14.1) at $z \sim 2.5$. Above $z \sim 2.5$,

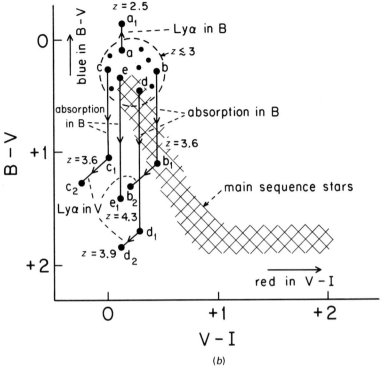

Fig. 14.4 (*Cont.*)

Lyα passes beyond the peak B-response and the B–R colours then progressively redden as first the Lyα forest absorption and then the Lyman limit cut-off reduce the flux in B. The reddening does not become really significant until the Lyman limit is well into the B-response at $z \sim 3$ so that typically all QSOs with $z < 3$ will lie in or just to the blue of the region of the broken circle. By $z \sim 3.5$, however, the combined reddening from the Lyα forest and the Lyman limit is of the order of 1 magnitude and continues to increase up to perhaps 1.5–2 mag at $z = 3.9$. Above $z \sim 3.9$ the situation becomes more complicated as Lyα moves into the R response. The increased flux in R not only enhances the reddening in B–R but simultaneously makes the QSO bluer in R–I by a similar amount. The result is that the QSO moves to the left in the (B–R, R–I) diagram, away from the main sequences at an angle of 45 deg; from the data in Table 14.1 the resultant shift for $W_R = 75\,\text{Å}$ is ~ 0.6 mag at $z \sim 4.3$. The solid lines in Fig. 14.4(*a*) show how, on these arguments, the QSOs 'a' to 'e' move in the (B–R, R–I) diagram as they are shifted to higher redshifts. As Lyα moves past the peak of the R-response and the Lyα forest absorption comes into play, a QSO will become both bluer in B–R and redder is R–I. It will therefore move back towards the main sequence stars and, as discussed below, the BRI technique will break down at $z \sim 4.5$, at least for the particular R-filter considered here. However, as R. McMahon has pointed out to me, at $z > 4.5$ those QSOs with a very weak continuum to the blue of the Lyman limit may be so red in B–R as the lie well below the horizontal section of the main sequence shown in Figs. 14.4 and 14.5. Therefore, they may still be easily detectable although possibly so faint in B that only an upper limit can be set to their B–R colour. How high a redshift can be achieved on the basis of the very red B–R colours will depend on the behaviour of the Lyα forest absorption at $z > 4.5$. Even a simple extrapolation suggests that this absorption could rise from 0.1 at $z \sim 2$ to ~ 1 mag at $z \sim 5$. Because the recognition of these very high redshift QSOs requires B–R > 2, the increased absorption in R implies that the BRI colours will be useful above $z \sim 4.5$ only for intrinsically bright QSOs near the upper end of the luminosity function (I $\lesssim 18$) and its success will depend on whether such bright QSOs exist in significant numbers at these high redshifts.

Figure 14.4(*b*) shows a schematic (B–V, V–I) diagram constructed in the same way as that in Fig. 14.4(*a*). The colour changes produced by the passage of the various spectral features through the B-response occur exactly as in the previous discussion. The difference is that Lyα emission comes into play at a lower redshift and has its maximum effect when it lies near the peak of the V-response at $z \sim 3.6$ and it is therefore the QSOs 'b' and 'c' which move diagonally away from the main sequence stars. The useful redshift limit for the BVI system is around $z \sim 4$ when Lyα lies near the red limit of the V-response and its contribution to the V magnitude will be almost exactly cancelled by the Lyα forest absorption. This accounts for the almost vertical displacement of QSO 'd'. The QSO 'e' will be displaced by a smaller amount in B–V than 'd' because Lyα now lies beyond the V-response and the QSO may in fact move back towards the main sequence stars.

As far as the basic selection effects are concerned the two colour (B–V, V–I) diagram of Fig. 14.4(*b*) is similar to the two colour (J–V, V–R) diagram used by Irwin *et al.* (1985) and Warren *et al.* (1987) and it reproduces the main features of their observations. Both 14.4(*a*) and (*b*) show clearly why high redshift QSOs tend to lie to the blue of the main sequence stars in V–I and R–I but, more importantly, they also illustrate how important to the selection process are (1) a weak continuum to the blue of Lyα (2) the presence of strong Lyα emission and (3) a blue continuum to the red of Lyα.

The ability to separate high redshift QSOs from stars depends on the QSOs being shifted from the upper left of the main sequence of the appropriate two colour diagram into a region where they are clear of the main sequence stars. That such a separation is possible relies on the correlated colour of the stars along the two axes which causes the main sequence to run more or less diagonally from top left to bottom right across the colour diagrams. A vertical downwards movement of a QSO with increased redshift can thus carry it from the top left of the diagram through and beyond the stellar population and the bigger the downwards movement the more likely it is to move into a sparsely populated region. For QSOs with weak Lyα emission the extent of the downwards movement depends mainly on the amount of the absorption in the B-band which reaches about 1.5–2 mag around $z \sim 4$. In principle, whether the other two passbands are V and R or R and I is largely irrelevant although the displacement in B–V is somewhat reduced relative to B–R because of more Lyα forest absorption in the V-filter. How well a QSO separates from the stars depends strongly on the colour of the continuum to the red of Lyα. The colour of this continuum is more critical for the smaller vertical displacements in B–R and B–V as can be seen by the considering the two QSOs 'b' and 'e'. While the blue continuum 'c' moves just to the left of the main sequence (c_1) and could possibly be separated by its colours, the redder continuum object 'b' moves into a hopelessly crowded region near the middle of the main sequence (b_1). Again the choice of passbands other than B is irrelevant and no permutation of B, V, R and I is of the any assistance. However, a narrower and bluer filter to replace the B-filter could have advantages since the Lyman limit absorption would then come into effect at a lower redshift and more sharply than in the broad B-filter. In principle this could be the U-filter, but a filter centered below 4000 Å offers no advantage with respect to the Lyman limit for $z > 3.3$ and, moreover, the stellar absorption below ~ 4000 Å could result in the QSO candidate list being more heavily contaminated with galactic stars.

For QSOs with strong Lyα emission lines the situation is completely different and the choice of the central filter becomes a prime consideration. A strong Lyα line moves a QSO away from the main sequence only if its redshift is such that Lyα falls within the passband of the central filter, at redshifts $z \sim 4.3 \pm 0.3$ using BRI or $z \sim 3.7 \pm 0.3$ for a survey using B, V and I. The role of Lyα is particularly significant at these lower redshifts. Around $z \sim 4$ the vertical displacement due to absorption in the B-band can be sufficient in itself to move a QSO clear of the main sequence stars and the effect of Lyα is a bonus which acts to enhance the separation. Around $z = 3.6$, however, the vertical displacement would still leave a QSO close to or within regions of high stellar density without the additional displacement of Lyα in the V-passband. The bias in the broad band colour surveys to QSOs with strong Lyα does not appear to have been widely recognized. Given this bias it is not surprising that the colour surveys in the SGP yielded no bright QSO with $z < 3.9$ which had not already been found in the IIIa–F prism surveys. The bias to strong emission line objects is small for $z \geqslant 4$. For colour surveys aimed at obtaining samples of high redshift QSOs over a wide range of emission line strengths there is therefore a case to be made for concentrating attention around $z \sim 4.3$ using B, R and I colours. At $z = 4.3$, QSOs with a very weak continuum to the blue of the Lyman limit will be displaced even further down the J–R axis than QSOs with $z \sim 4$. In principle, this makes them easier to detect but they may then be so faint in B that accurate magnitude measurements will not be possible. This in itself may not be a serious limitation but implies that the initial reference survey must be carried out in R or I and not B.

As has already been pointed out the (J–V, V–R) diagrams shown by Irwin *et al.* (1985) and Warren *et al.* (1987) are essentially equivalent to the schematic (B–V, V–I) diagram of Fig. 4(*b*). Apart from a shrinkage of the horizontal scale for (J–V, V–R) the only significant difference with (B–V, V–I) occurs because of an overlap in the passbands of the V and R filters. As a result of this overlap, at redshifts above $z = 3.9$ the Lyα emission in R counteracts its contribution in V and with increasing redshift above $z = 3.9$ results in the QSO becoming redder in V–R. This accelerates the breakdown of the colour selection as Lyα moves out of the V-filter and which we have already discussed for the BVI and BRI systems. The (J–V, R–I) plot shown by Warren *et al.* (1987) is of interest because it confirms that Lyα emission does indeed play a significant role in separating QSOs from the main sequence stars. Replacing V–R by R–I removes the displacement along the horizontal axis produced by Lyα in the V filter. The $z = 3.61$ and 3.67 QSOs would, therefore, be expected to lie closer to the main sequence stars in (J–V, R–I) than in (J–V, V–R). On the other hand, for the $z = 4.01$ QSO, Lyα in the R filter would be expected to move it slightly further from the stars in (J–V, R–I). The observed colour diagrams confirm that the expected displacements do occur.

33(c) Medium band (200–300 Å) filter surveys

When considering in 1973 the colour selection effects in UBV surveys, and particularly the colour changes produced by emission lines in the narrow U-filter, it occurred to me that rather than considering the colour changes produced by the emission lines as a problem they could be used themselves as the basis of a powerful selection procedure. One obvious way of maximizing the effect of the emission lines was to use a set of filters matched to the widths of the lines (Hazard, 1980). The method appeared ideally suited to surveys for high redshift QSOs by their Lyα emission. Thus, consider a QSO with a strong Lyα emission line with $W_R = 200$ Å. At a redshift $z = 4$ the observed equivalent width will be $W = 1000$ Å while the width of the line will typically be about 200 Å. In a filter with a passband 200 Å wide such a line will produce a brightening of about 2 magnitudes which can easily be detected. Even a QSO with W_R as small as 25 Å produces a brightening of ~ 0.5 mag and moreover the equivalent width increases with redshift. In collaboration with N. Sanitt I also considered the alternative of using an objective prism on a Schmidt telescope to detect the emission lines. However, we based our calculations on the performance of the emulsions used in the Palomar Sky Survey and concluded that such a system would work well to about magnitude $+ 18$ but that to reach magnitude $+ 20$ would mean restricting the bandwidth to around 600 Å or less (Sanitt, 1974). This is too narrow to be used effectively with the broad QSO lines and the dispersion of the prism scarcely comes into play. This analysis, therefore, again indicated a survey based on medium band filters alone. As a result I started a pilot search for high redshift QSOs on the UKST using a set of five interference filters with 8 in diameters (about 12 deg² of sky). In the converging beam of the UKST each filter has a half-power bandwidth (HPBW) at the 50% power level of about 250 Å and their nominal central wavelengths are separated by 200 Å. They cover a wavelength interval of 5600–6600 Å which for Lyα corresponds to $3.6 < z < 4.4$. However, there is a shift to shorter wavelengths and to broader bandwidths towards the edge of the field because of the increasing angle of incidence.

As it turned out, our estimate of the performance of objective prism surveys was for too pessimistic, having failed to take into account the improvement in photographic emulsions since the POSS had been completed. When the true power of the technique had been demonstrated by Smith (1975), and the high quality of the UKST material become apparent, my attention was diverted more to objective prism surveys. There were a few other reasons for delaying the programme. Firstly, the analysis techniques for the then new APM had not at that time reached the stage where the filter data could be properly exploited. Secondly, there were then no QSOs known where, using that particular set of filters, the technique could be tested on

Lyα, although preliminary tests did show that CIV could be detected in redshift $z \sim 3$ QSOs. Finally, the basic idea was to construct very low resolution spectra between 5600 and 6600 Å and obtaining a set of five high quality deep plates in any particular field proved more difficult than expected. Also we were having a fair degree of success with the IIIa–F prism surveys which left little time for the filter work. The above problems have now been overcome. The results of the broad band colour surveys have shown that the APM machine and its associated software have reached a high level of perfection, bright calibration QSOs are available and, more importantly, the analysis presented in the previous section indicates how the calibration of the filter plates can be conveniently carried out.

3(c) (1) A test of the medium band filters

Over the years we have accumulated a fairly large collection of filter plates which are being analyzed. These include a set covering the central region of the SGP which contains the $z = 3.61$ QSO DHM0054–284 and the $z = 3.67$ QSO 0055–2659. I describe here a test of the filter technique based on the detection of Lyα from these two QSOs in the lowest wavelength filter of the set of five. The results of this test indicate that the central wavelength of this filter in the converging beam of the UKST probably averages around $\lambda \sim 5700$ Å and is thus well suited to the detection of QSOs with $z = 3.6$. The simplest method of analysis requires a single comparison plate which must be insensitive to Lyα at $z \sim 3.6$. This could be another medium band filter plate or a plate taken using a very broad filter which should have a central wavelength reasonably close to $\lambda \sim 5700$ Å. The signature of a QSO will then be an image which is significantly brighter in the 5700 Å filter plate than in the comparison plate while the colours of other objects in the two plates will be approximately zero. An alternative procedure is to follow the method used for the broad band colour surveys. The analysis presented in the previous section shows that the two colour JVR diagram relies for much of its success on Lyα lying in the V-passband which suggests using a similar two colour diagram for the filter survey by replacing the V-magnitudes with magnitudes (X) from the filter plate. I describe an analysis used by R.G. McMahon and myself which combines both the above methods. It uses UKST R-plates as a broad band reference and two colour diagrams in J, X and I and J, R and I.

Figure 14.5 shows a two colour (J–R, R–I) diagram covering the whole of

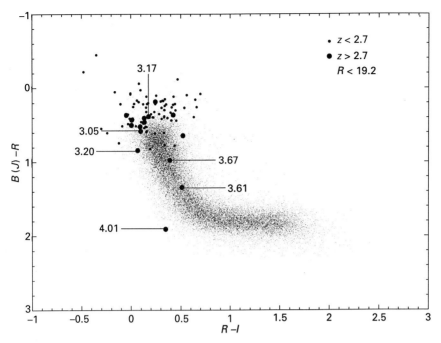

Fig. 14.5. Observed (J–R, R–I) diagram for the SGP field centred on RA 0053, Dec –2803, showing known objects with R < 19.2 in different redshift intervals. Objects with $z > 3$ are shown individually and indicate the displacement with increasing redshift to redder values of J–R; the $z = 3.67$ QSO lies higher than might be expected because of an unusually strong blue continuum. Compare with Fig. 14.4(a).

the SGP field which shows the $z = 4.01$ QSO discovered by Warren *et al.* (1987), the $z = 3.61$ and 3.67 QSOs and several lower-redshift QSOs mostly selected from UKST IIIa–J prism plates. It illustrates the main features of the schematic representation presented in Fig. 14.4(a), notably (1) the QSOs with $z < 3$ clustered at the top left of the diagram just above the main sequence, (2) the movement down the J–R axis with increasing redshift, and (3) the clear separation of the $z = 4.01$ QSO produced by the redward displacement of the main sequence stars with increasing J–R and which was so successfully exploited by Warren *et al.* (1987) and McMahon *et al.*, (1988) in the discovery of the $z = 4.01$ and $z = 3.97$ QSOs. It is noteworthy that the confirmed low redshift QSOs already comprise most of the objects with J–R < 0.3. This suggests that, with the accuracy which can be obtained with the APM, a single colour selection based on J–R could be competitive with U–B selection and moreover extend the redshift range to $z \sim 3$. The wide spread of these low redshift QSOs in R–I is also significant because it indicates that at large redshifts, absorption effects would move many of them down into, or to the right of, the main sequence and hence, only detectable if they have very strong Lyα emission. The diagram also illustrates why a colour comparison depending only on R–I is a useful check of IIIa–F objective prism high redshift candidates. The prism selection ensures that all such candidates are very red in B–R so that the colour check is needed only to show that they have R–I < 0.8. Fig. 14.6(*a*) shows the same two colour plot but for only the central region of the SGP covered by the 5700 Å interference, filter, the X-filter. Fig. 14.6(*b*) shows the (J–X, X–I) diagram which results when the R-filter is substituted by the X-filter. Lyα emission from the $z = 3.61$ and $z = 3.67$ QSOs in the X-filter now results in these QSOs becoming simultaneously redder along the ordinate and bluer along the abscissa so that they move away from the main sequence at an angle of 45 deg. This is exactly the same as occurs on a (J–V, V–R), or (J–V, V–I) diagram but because of the narrower X-filter the displacement is significantly enhanced. The $z = 3.61$ and 3.67 QSOs have equivalent widths of Lyα in the observed frames of about 275 and 540 Å, respectively. The corresponding magnitude changes in the 1300 Å wide V-filter are 0.2 and 0.4 mag which produce displacements from the main sequence of ~ 0.3 and ~ 0.6 mag respectively, in good agreement with the values observed by Warren *et al.* (1987). In the narrow X-filter the expected magnitude changes in X for $\Delta\lambda = 250$ Å are 0.8 mag and 1.2 mag corresponding to maximum displacements of 1.1 and 1.7 respectively. The observed values are ~ 0.9 and 1.0. The smaller than expected displacement for the $z = 3.67$ QSO suggests it is not so well centred in the filter as the $z = 3.61$ QSO, probably due to it lying near the edges of the field where the filter wavelength is shorter and the bandpass larger. A single filter does not provide a full test of the medium band filter technique because observations over the full set are required to ensure that the contribution from the line is maximized in one of them. Nevertheless, the results from this single filter confirm that the medium band filters provide a powerful technique for detecting even relatively weak Lyα emission.

The medium passband filters are sensitive not only to Lyα emission but also to weaker lines in the spectrum of QSOs at lower redshifts. These lower redshift QSOs are easily recognized from their positions on the (J–X, X–I) and (J–R, R–I) diagrams. Object 'c' in Fig. 14.6 is an example of a low redshift QSO with a possible emission line in the X-filter. Its IIIa–J objective prism spectrum shows a QSO with $z < 2$ and an emission line at $\lambda \sim 3760$ Å. One interpretation would be that the line in the prism spectrum is CIII] at

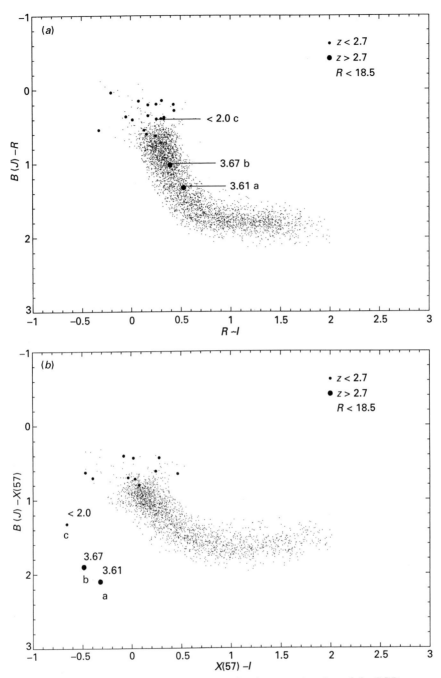

Fig. 14.6. Observed (J–R, R–I) diagram for the central region of the SGP compared with a (J–X, X–I) diagram for the same region, where X represents a narrow band filter centred on $\lambda \approx 5700$ Å. 'a', 'b' and 'c' are QSOs which are displaced significantly between the two diagrams. QSOs 'a' and 'b' are displaced because of Lyα in the narrow X-filter. QSO 'c' which moves from the $z < 2$ region of 14.5(a) is a QSO with $z \leqslant 2$ which may have strong MgII in the X-filter but more probably has varied in brightness between taking the X-filter and taking the R and I plates. Note the large displacement of 'a' and 'b' off the main sequence only.

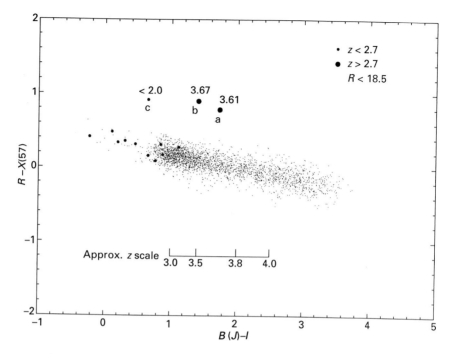

Fig. 14.7.   (R–X, J–I) diagram for the central region of the SGP obtained by taking the difference between Fig. 14.6(a) and 14.6(b). Note that QSO 'c' which moves across the main sequence between Fig. 14.6(a) and 14.6(b) is more clearly separated from the main sequence stars in this diagram. The J–I axis is a rough redshift axis. Genuine $z \sim 3.7$ QSOs for the 5700 Å filter should lie off the main sequence only in the interval $1.2 < J–I < 1.8$.

$z \sim 1$ which corresponds to MgII at $\lambda \sim 5600$ Å. However, the displacement between the two diagrams seems to large to be explained by MgII and it is more probable that it arises because the QSO has varied in brightness in the interval between taking the filter plate and taking the J and I plates.

Object 'c' was recognized as a QSO not so much by its position off the main sequence in Fig. 14.6(b) as by its large displacement between 14.6(a) and 14.6(b). The displacement of a QSO by one of its emission lines falling in the X-filter is as much a key to its recognition as where it finally ends up in the (J–X, X–R) diagrams. A QSO with a very red R–I colour could obviously move a considerable distance between the two diagrams only to end up in the middle of the main sequence. This suggests that a plot showing the displacement between the position of all corresponding points in the two diagrams is a more useful presentation of the data. A little bit of algebra shows that this is simply equivalent to measuring the colour index (R–X), that is the difference in magnitude between the broad and narrow band filters. This is precisely the method suggested at the beginning of this discussion. However, the value of obtaining the difference through the pair of two colour diagrams is that it allows the variation in the colours of stars between the two passbands to be taken into account. Figure 14.7 shows the final result of taking the suitably corrected difference between Fig. 14.6(*a*) and (*b*) where R–X has been plotted against J–I. It follows from the previous discussion that J–I is essentially a redshift scale which has been indicated on the diagram; the $z = 3.67$ QSO is bluer in J–I than would be expected but it is known to have an unusually strong continuum to the blue of Lyα with no cut-off near the Lyman limit. The displacement of an object off the main sequence at the appropriate value of J–I for Lyα in the particular filter employed is a useful confirmation that it really is a high redshift QSO. For example with the 5700 Å filter no genuine objects should appear with (J–I) > 2 while all objects with (J–I) < 1.2 are probably QSOs at lower redshifts. With the full set of filters, genuine QSOs should gradually move to redder values of (J–I) as diagrams are constructed corresponding to successively larger redshifts.

The (R–X, J–I) diagrams are a useful way of presenting the filter data. However, the selection of QSOs then relies entirely on emission lines in the X-filters. Apart from the rough redshift information in the J–I scale no use is made whatsoever of the absorption in the Lyα forest and/or at the Lyman limit as a possible selection criterion. This seems a pity considering that the absorption produces a displacement in J–R of the order of two magnitudes around $z \sim 4$. Even the detection of the known $z \sim 4$ QSOs used only the displacement the absorption produces relative to the main sequences stars (~ 0.5 mag) rather than the total displacement. It could perhaps be useful to replace the reference two colour (J–R, R–I) diagram with a reference plot where J is replaced with a filter to the red of the expected position of the Lyman limit. There would certainly be definite advantages in using a narrower and bluer filter than either J or B so that the vertical displacement in J–R or J–X occurs more sharply and at lower redshifts. A relatively narrow filter around $\lambda \sim 4200$ Å (say Y) would bring QSOs with redshifts as low as 3.6 into the same position in the (Y–R, R–I) diagram as $z \sim 4$ QSOs on a (J–R, R–I) diagram. The heavy absorption in steller spectra at $\lambda \leqslant 4000$ Å tends to reduce the effectiveness of a U-filter for this purpose.

4 Discussion

The main problem in searches for very high redshift QSOs is posed by their very low surface density which necessitates that the surveys be carried out over very large areas of sky. This in turn imposes very stringent limits on the quality of the data, the selection criteria and the data analysis; at redshifts above $z > 4$ it may be required to detect a single high redshift QSO from among perhaps 10^5 possible contaminating stars. That eight of the QSOs with $z > 3.5$ listed in Table 14.2 (and eight of the nine optically selected objects) should have been found using UKST material is therefore both an indication of the importance of Schmidt telescopes in extragalactic research and a tribute to the quality of the telescope and the work of the UKST group. The recent success at $z \sim 4$ (Warren, *et al.*, 1987; McMahon *et al.*, 1988) using the APM demonstrates the impact which the new measuring machines are now having in large scale survey work and which can be expected to grow rapidly in the future.

The three optical search techniques using (1) IIIa-F objective prism plates, (2) broad band colours, or (3) narrow band filters each have their particular advantages and disadvantages. Each is still in an early stage of development and there is as yet insufficient observational data to compare them directly. However, the analysis presented earlier permits some general conclusions on how they may best be used. An important consideration in any such discussion is how easy or difficult it might be to obtain the required Schmidt plates.

First consider a IIIa-F objective prism survey and a broad band colour survey based on J, V, R and I which will be assumed to be carried out in a region where the standard survey plates in J, R, and I are already available. Each survey has the same upper redshift limit which is set by the red cut-off of the IIIa-F emulsion. As we have described, the JVR system should be used up to $z \sim 3.9$-4.0 while, above this redshift, JRI becomes more suitable. The JVR survey thus covers roughly the same redshift interval as the IIIa-F prism plates taken using the low dispersion objective prism on the UKST (if we assume the lowest redshift of interest is $z \sim 3.3$) and, in this interval, both techniques are biassed to objects with strong Lyα emission. For bright QSOs ($R < 18.5$) the data collection problems are comparable, the prism survey requiring one 20 min exposure unfiltered IIIa-F plate while sufficiently accurate broad band magnitudes can probably be obtained by taking a 30–60 min V-plate to supplement the general survey plates. The prism plate, however, has the advantage of being able to yield highly reliable selections with relatively accurate redshifts estimated directly from the prism spectra while the colour survey yields only a list of QSO candidates. In surveys for fainter QSOs the balance tilts towards the JVI survey which can probably reach to magnitude $\sim +20$ but at the expense of at least 3 additional plates, one each in J,V and I, to ensure the necessary photometric accuracy.

However, the addition of a filter to limit the prism spectra in the blue at $\lambda \sim 4950\,\text{Å}$ can reduce the limiting magnitude in prism surveys to $R \sim$ 19.5 mag so that the colours may not have an overwhelming advantage. The real advantage of the broad band colours occurs at $z > 3.9$ when Lyα near the head of the prism spectra becomes difficult to separate from features in the spectra of cool stars. On good quality prism spectra, strong lines can be recognized up to $z \sim 4.2$ but broad band colours, with additional plates as before, should be able to reach to magnitude ~ 20 and to $z \sim 4.6$. Moreover, at redshifts $z > 4$ the colour surveys become much less dependent on the presence of strong Lyα emission. However, little work has yet been done using filtered prism plates taken with the 21/4 deg prism at the UKST which has a dispersion nearly three times that of the low dispersion prism. With the spectrum limited in the blue to $\lambda = 5900\,\text{Å}$ there seems no reason why such plates should not be viable up to $z \sim 4.5$ and to magnitude ~ 19 for at least strong emission line QSOs.

If two plates are required in each passband, a deep broad band colour survey covering the interval $3.3 < z < 4.6$ will require taking at least seven new deep plates in each survey region. Good high quality deep prism plates are also very difficult to obtain. Therefore, it is unlikely that deep surveys can be carried out in more than a few regions of sky, unless a large amount of Schmidt telescope time is specifically directed to the project. Surveys over large areas will in general have to be restricted to the brighter QSOs although in BRI, QSOs around $z \sim 4$ may be well enough separated from galactic stars to allow the selection of even relatively faint QSOs with only a single plate in each passband (McMahon *et al.*, 1988). Surveys in BVI using one plate in each passband have the great advantage that they can be based entirely on survey material in UKST plate library. However, there should be no serious problem in obtaining enough unfiltered IIIa-F prism plates or direct V-plates to also survey the interval $3.3. < z < 3.9$ over large areas.

For deep surveys in a few regions, or surveys for bright QSOs over large areas of sky, there are advantages in combining the broad band colours and the objective prism data. With the help of the objective prism spectra at least a large fraction of the contaminating stars and emission line galaxies can be removed from the colour selected candidates thus reducing the number of objects which need to be followed up on large telescopes. This in turn allows a relaxation of the colour selection criteria and permits the selection of objects closer to the main sequence in the two colour diagrams than might otherwise be possible. Alternatively, the broad band colours can be used to check objects selected from the prism plates. Because the automatic measuring machines are so well suited to the analysis of the colour plates it might be argued that the optimum procedure is always to select first by colour and then examine the prism spectra of the colour selected candidates, either by eye or using the measuring machine. However, it has been shown (Hazard *et al.*, 1987) that the prism plates are capable of yielding not only strong emission line QSOs but also high redshift BAL QSOs which in general will be missed in the colour surveys. Eye searches of the unfiltered IIIa-F plates, are not, in fact, particularly onerous taking only a few hours per plate.

The narrow band filters have the advantage of being able to detect QSOs with weaker emission lines than either broad band colour or the objective prism techniques. As even lines with equivalent widths in the rest frame as low as 50 Å produce a displacement from the main sequence stars of ~ 1 mag the accuracy of the magnitude measurements is less critical than for the broad band colours and one plate taken with each filter should be sufficient. However, this is achieved at the expense of a smaller redshift range per filter and to reach close to magnitude 20, will require exposures of the order of 2 hours per filter. However, these exposures can be made in grey time where there is less demand for telescope time. The optimum bandwidth increases with increasing redshift so that the filters become particularly valuable at $z > 4$ where the low dispersion prism plates have limited value. The existing filters cover only about one third the area of the field of view of the UKST but a new set is now being constructed with bandwidths of ~ 300 Å which utilize the whole field of view and will cover the wavelength interval 5800–6900 Å or $3.8 < z < 4.6$. They are a mixture of interference filters and Schott glass filters used with different emulsions. It is worth noting that the effective passband of the Palomar E-plates is only about 450 Å wide and that Lyα with $W_R = 75$ Å will produce in them a brightening of about 0.7 mag at $z \sim 4.3$ (see Table 14.1). An analysis based on the Palomar survey could therefore, be useful in searches for QSOs around this redshift. Variability in brightness since the epoch of the POSS is unlikely to be a serious problem. On the average, QSO variability is as likely to help in the selection of high redshift QSOs as to discriminate against them and our work up to now suggests that contamination from variable stars is not large.

How much effort should be expended in high redshift searches is a different matter, particularly at redshifts $z \gg 4$ where an extrapolation of the observed decline in numbers from $z \approx 2$ suggests they may be very rare objects indeed. The steady decline with increasing redshift indicates that QSOs are being born over the whole redshift range above $z \sim 2$ so that it is unnecessary to go to very high redshifts to study them in the early phases of their evolution. It is possible that the QSO luminosity function extends to higher luminosities at the higher redshifts but otherwise the properties of the presently known QSOs with $z \sim 3.5$ do not appear to differ significantly from those of QSOs around $z \sim 2$. This indicates that perhaps their major role will be in absorption line studies of the intervening intergalactic medium. This in turn indicates that searches for bright examples over large areas of sky should have a high priority. In any case, as the luminosity function at high redshifts is relatively flat, the increase in the number of QSOs which will be detected by extending the survey limit from magnitude $+18.5$ to magnitude $+20$ will be relatively small. Therefore, surveys over large areas even if only to a moderate limiting magnitude may not only be easier but more productive than deep surveys over smaller areas of sky (Hazard & McMahon, 1985). Deep surveys over a few areas are, of course, required to better define the luminosity function. However, for the moment, it might be better to concentrate such investigations at lower redshifts, say around $z \sim 3$, where the QSO density is still relatively high. There is, at present, no reliable information on the luminosity function even at this relatively low redshift.

The usefulness of the various search techniques does not depend solely on their use in high redshift QSO surveys. Each technique has far wider applications. A low dispersion IIIa-F prism survey using unfiltered plates also yields about 200 lower redshift QSOs per Schmidt field as well as samples of BAL QSOs and interesting stars such as C-stars. The broad band colour technique used by Warren *et al.* (1987) also uses plates taken with a U-filter so that it is also not confined to high redshift QSOs but yields also large samples at $z \leqslant 2.2$. Perhaps the main use of these accurate broad band surveys will be in the study of stellar distributions and in searches for unusual types of stars. Rather than searching for QSOs with redshifts well in excess of $z \sim 4$, where there may be only one such QSO per several Schmidt plates, it may be more productive to concentrate on these other investigations and hope that the high redshift QSOs turn up as a by-product of these studies. The narrow band filters could be particularly valuable in setting limits to the surface density of very high redshift objects. In a $300 \text{\AA}$ filter, QSOs with line strengths similar to those of the strongest emission line QSOs already known with $3.5 < z < 4.0$, will lie off the main sequence in the two colour diagram by between 1.4 and 2 mag at $z = 5$. This is such a large displacement that even a null result in a filter survey over only a few Schmidt fields would set a quite definite and significant limit to the surface density at $z \sim 5$. It would show a much lower surface density than around $3.5 < z < 4.0$ or alternatively show that the QSOs were no longer characterized by strong Lyα emission. Using the analysis techniques described above a survey could be carried out using only a single narrow band filter.

Acknowledgements

I thank the members of the UKST unit who have provided the objective prism and filter plates for the high redshift surveys. The analysis of the filter is a collaborative project with R.G. McMahon and M.J. Irwin of the APM group at the Institute of Astronomy, Cambridge. I particularly acknowledge the contribution of R.G. McMahon and the extensive discussions I have had with him on the selection procedures, and also thank him for preparing the colour-colour diagrams presented in Figs. 14.4 to 14.7.

References

Anderson, S.F. & Margon, B. (1987), *Nature*, In press.

Bahcall, J.N. & Sargent, W.L.W. (1967). *Ap. J., (letters)*, **148**, L65.

Baldwin, J.A., Smith, H.E., Burbidge, E.M., Hazard, C., Murdoch, H.S. & Jauncey, D.L. (1976). *Astrophysics J. (letters)*, **206**, L83.

Beaver, E.A., Harms, R., Hazard, C., Murdoch, H.S., Carswell, R.F. & Strittmatter, P.A. (1976). *Ap. J. (letters)*, **203**, L5.

Carswell, R.F. & Strittmatter, P.A. (1973). *Nature*, **262**, 394.

Clowes, R.G., Cooke, J.A. & Beard, S.M. (1984). *Mon. Not. R. Astr. Soc.*, **207**, 99.

Dunlop, J.S., Downes, A.J.B., Peacock, J.A., Savage, A., Lilly, S.J., Watson, S.G. & Longair, M.S. (1986). *Nature*, **303**, 564.

Gearhart, M.J., Lundt, J.M., Frantz, D.J. & Kraus, J.P. (1972). *A. J.*, **77**, 557.

Gent, H., Crowther, J.H., Adgie, R.L., Hoskins, D.G., Murdoch, H.S., Hazard, C. & Jauncey, D.L. (1973). *Nature*, **241**, 261.

Hayman, P.G., Hazard, C. & Sanitt, N. (1979). *Mon. Not. R. Astr. Soc.*, **189**, 853.

Hazard, C. (1973). *Nature*, **242**, 365.

Hazard, C. (1977a). In *Radio-astronomy and Cosmology* (ed. G.L. Jauncey), Dordrecht: D. Reidel, pp. 157.

Hazard, C. (1977b). In *Quasars and Active Galactic Nuclei* (eds. C. Hazard and S. Mitton), Cambridge University Press, p. 1.

Hazard, C. (1980). In *Variability in Stars and Galaxies*, (Liege: Institut Astrophysique).

Hazard C. (1985). In *Active Galactic* (ed. J. Dyson), Manchester University Press, p. 1.

Hazard, C. & McMahon, R.G. (1985). *Nature,* **134**, 238.

Hazard, C. & McMahon, R.G. (1987). In preparation.

Hazard, C., McMahon, R.G. & Morton, D.C. (1987). *Mon. Not. R. Astr. Soc.* Submitted.

Hazard, C., McMahon, R.G. & Sargent, W.L.W. (1986a). *Nature*, **38**, 322.

Hazard, C., Morton, D.C., McMahon, R.G., Sargent, W.L.W. & Terlevich, R. (1986b). *Mon. Not. R. Astr. Soc.*, **223**, 87.

Hazard, C., Terlevich, R., McMahon, R.G., Turnshek, D., Foltz, C., Stocke, J. & Weymann, R. (1984). *Mon. Not. R. Astr. Soc.*, **211**, 45pp.

Hewett, P.C. *et al.* (1985). *Mon. Not. R. Astr. Soc.*, **213**, 971.

Hewitt, A. & Burbidge, G. (1987). *Ap. J. Suppl.*, **63**, (1).

Hoag, A.A. & Smith, M.G. (1977). *Ap. J.*, **217**, 362.

Irwin, M.J., McMahon, R.G. & Hewett, P.C. (1985). *Measuring Machines Newsletters*, no. 8, SERC.

Kibblewhite, E.J., Bridgeland, M.T., Bunclark, P.S. & Irwin, M.J. (1983). *Proc. Astronomical Microdensitometry Conf. 277*, NASA, Washington, D.C.

Koo, D.C. (1983). *24th Liege. Int. Astrophys. Coll.* **240.**

Koo, D.C., Kron, R.G. & Cudeworth, K.M. (1986). *P.A.S.P.*, **98**, 285.

Landau, R. & Ghigo, F.D. (1983). *Proc. Astronomical Microdensitometry Conf. 277* NASA Washington D.C.

Lynds, C.R. & Wills, D. (1970). *Nature*, **226**, 532.

Markarian, B.E. (1967). *Astrophyzika*, **3**, 55.

Markarian, B.E., Lipovetskii, V.A. & Stepanion, D.A. (1979). *Astrophyzika*, **15**, 549.

Marshall, H.L., Avri, Y., Bracessi, A., Huckra, J.P., Tananbaum, H., Zamorani, G. & Zitelli, V. (1983). *Ap. J.*, **369**, 352.

McGillivray, H.T. & Stobie, R.S. (1985). *Vistas in Astronomy*, **27**, 433.

McMahon, R.G., Irwin, M.J. & Hazard, C. (1988). In preparation.

Nandy, K. Reddish, V.C., Tritton, K.P., Cooke, J.A. & Emerson, D. (1977). *Mon. Not. R. Astr. Soc.*, **178**, 63 pp.

Osmer, P.S. (1982). *Ap. J.*, **253**, 28.

Osmer, P.S. & Smith, M.G. (1976). *Ap. J.*, **210**, 267.

Peterson, B., Savage, A., Jauncey, D.L. & Wright, A.E. (1983). *Ap. J.*, **260**, L27.

Sandage, A.R. (1965). *Ap. J.*, 141, 1560.

Sanitt, N. (1974). PhD Thesis, Cambridge University.

Schmidt, M. (1968). *Ap. J.*, **151**, 393.

Schmidt, M. & Green, R.F. (1983). *Ap. J.*, **369**, 352.

Schmidt, M., Schneider, D.P. & Gunn, J.E. (1986). *Ap. J.*, **310**, 518.

Shanks, T., Fong, R. & Boyle, B.J. (1983). *Nature*, **303**, 156.

Smith, M.G. (1975). *Ap. J.*, **202**, 591.

Smith, M.G., Boksenberg, A., Carswell, R.F. & Whelan, J.A.J. (1977). *Mon. Not. R. Astr. Soc.*, **181**, 67 pp.

Strittmatter, P.A. & Burbidge, G.R. (1967). *Ap. J.*, **163**, 13.

Veron, M.P. (1971). *Astron. Astrophysics*, **11**, 1.

Wampler, E.J., Robinson, I.B., Baldwin, J.A. & Burbidge, E.M. (1973). *Nature*, **263**, 336.

Warren, S.J., Hewett, P.C., Irwin, M.J., McMahon, R.G., Bridgeland, M.T., Bunclark, P.S. & Kibblewhite, E.J. (1987). *Nature*, **325**, 131.

Watson, K. (1986). Preprint.

15

The ESO 16 m VLT

L. WOLTJER

Summary

The ESO 16 m Very Large Telescope (VLT) is an array of 4 telescopes, each of diameter 8 m. Due to be completed around 1998, it will have a light-gathering power 10 times that of the Hale 200 in reflector.

Major progress in science has frequently resulted from an order of magnitude improvement in some characteristic of the instrumentation. Following the completion of the 200 in telescope at Mt Palomar, large improvements in the light-collecting power of telescopes have come about from the introduction of detectors of high quantum efficiency, such as CCDs. Since some of these reach efficiencies of 60% and more, it is clear that further progress in this area will have to come in other ways, more specifically from an enlargement of the collecting area. There are no major technical obstacles to achieving this. However, the cost of larger telescopes is so high that their realization can only be assured by new technologies which allow very substantial cost reduction. The task of the modern telescope designer is not necessarily to make the largest telescope, but rather the best telescope that we can afford to build.

At ESO, plans have been made for a telescope with a total aperture of 16 m, which would result in a light-gathering power ten times larger than that of the 200 in (see *ESO Proceedings* 24, and Woltjer, 1984). The fundamental problem is how to make the primary mirror for such a telescope. A 16 m monolithic mirror is not a possibility. Even if we could manufacture it and pay for it, we would have the greatest difficulties transporting it. So we have to subdivide the aperture, and there are three ways to do this.

In the segmented mirror approach, a large mirror is made out of smaller pieces which are appropriately polished and then oriented in the right direction. This will be done in the California 10 m Keck telescope project. Problems include the difficulty of avoiding edge-effects between the segments and the complexity of the control system needed to maintain the overall shape to a fraction of a wavelength.

In the multi-mirror telescope, the situation is simpler. Several monoliths are polished and placed in a common mount and the light combined by some appropriate optics. The support and control problems are simpler but, of course, the diffraction pattern is different from that of a single dish and light losses are incurred in the beam combination.

The situation is rather similar in the array of independent telescopes. Instead of putting the various monoliths in a single mount, the mounts, too, are separated. The light may again be brought to a common focus, but the combining optics tend to be more complex. At ESO, this option has been chosen because of its flexibility: the telescopes may be used in a combined mode as a 16 m telescope, and also separately. Thereby, one eliminates the large gap between the presently available 4 m telescopes and the 16 m telescope.

If an array is chosen, how large should the individual telescopes be made? If they are chosen too small, there has to be a large number and the beam combination becomes complex and expensive, and the cost of instrumentation at the individual telescopes also becomes large. If they are chosen too large, the manufacture and transport of the mirrors becomes difficult and costly; quantitatively, this means that the individual telescopes should have diameters less than something like 8–10 m. They should, however, not be much less than this also for scientific reasons: an 8 m telescope has, at a wavelength of 20 μm, a diffraction disk of about 0.5 arcsec. At a good site, atmospheric conditions should allow observation with this type of resolution. As a result of such considerations, the diameter of the unit telescopes was fixed at 8 m and their number, therefore, at 4.

For the mirror blanks, several materials may be considered: glass, fused silica, Zerodur, nickel coated aluminum and steel. Experiments with the last two, which are relatively cheap, are being made at ESO; the main problem is to achieve a sufficient long term stability. An advantage of metals is their good thermal conductivity, a disadvantage is their large expansion coefficient—which is shared by glass. Silica and Zerodur are low expansion materials, easy to polish, but relatively expensive.

To reduce the cost of the mirrors, to improve their thermal time constant, and especially to reduce their weight, the mirrors should be thin (15–20 cm for an 8 m meniscus). With a purely passive mirror cell, such mirrors cannot be adequately supported against gravity; active optics is a necessity. In an active optics scheme, the image of some relatively bright star in the field is monitored and the necessary corrections to the mirror shape determined; the (motorized) supports are then adjusted to obtain the correct shape for the mirror. Laboratory measurements at ESO on a 1 m diameter mirror with a thickness of 25 mm supported by some 70 motorized levers have shown the feasibility of this approach.

The primary mirror should not only be thin, but also fast. If a large focal length were chosen, the telescope tube becomes long and heavy and the whole structure expensive. But a very fast mirror makes the optics more complex. At ESO, an $f/1.8$ primary has been chosen as a compromise. With a short, stiff tube the resonant frequencies of the telescope can be pushed up to near 10 Hz.

Conventional telescope buildings are expensive. At La Silla, the 3.6 m telescope is mounted 30 m high inside a dome with a diameter of 30 m. The cost of the building and dome was very close to that of the telescope itself! Such big domes are not necessarily creating optimal observing conditions. They may contain a large volume of relatively warm air which convects out slowly during the night, causing poor seeing conditions. At ESO, the possibility is being explored to have no dome at all, to have the telescope operate in the open air with only a wind screen if the wind is strong. Of course, the telescope has to be covered during the daytime and in bad weather. A sort of roll-off roof might be suitable for this, but at ESO also experiments are to be conducted with inflatable shelters.

An array has one important fringe benefit. It can also be used for interferometry. Especially in the infrared where the tolerances are easier to meet, interferometric use of the VLT is foreseen. The size of the seeing disk is of much importance for interferometry. To optimize the situation, adaptive optics (atmospheric compensation) will have an important role. In addition, a site selection campaign is under way to look for a place with minimal atmospheric turbulence. Moreover, it is important for the infrared work to have a site with a very low atmospheric water vapour content. In the ESO site survey, very dry sites have been found in the north of Chile.

The VLT will have unsurpassed light collecting power, combined with high angular resolution. A decision of the eight ESO member-countries on its construction is expected late in 1987. If so, the first 8 m unit telescope should be ready in 1993 or 1994 and the whole array in 1998. The cost is estimated at about 100000000 pounds.

A wide variety of studies will become possible with the VLT: High spectral resolution studies of intergalactic matter in front of quasars, studies of the nature of galaxies at redshifts of order unity and beyond, studies of the synthesis of the elements in our and other galaxies, very detailed quantitative studies of nuclei of galaxies, and studies of the processes of the formation of stars and of solar system like structures are just a few of the many examples that come to mind. And, of course, unexpected discoveries may have an even greater impact. Eight European countries are about to embark on their project of the future in optical astronomy. Does the UK really want to remain on the sidelines?

References

Proceedings 'Second Workshop on ESO's Very Large Telescope' (1986). *ESO Proceedings* 24.

Woltjer, L. (1984). In *Frontiers of Astronomy and Astrophysics*, ed. R. Pallavicini, Italian Astron. Soc., p. 3.

16

The Hubble Space Telescope

GARTH ILLINGWORTH

Summary

The scientific role that the Hubble Space Telescope, HST, will play in an era of 8–15 m ground-based telescopes is discussed. An overview of the spacecraft is given, with emphasis on the pointing and control system. A summary is made of some of the major tests that have been carried out, with an assessment of the expected performance and the status of HST as of early 1987. An overview of the ground- or operational systems is given, with particular emphasis on the characteristics of the system that make its operation so complex.

1 Scientific case

HST will be flown in an era of ground-based 8–15 m telescopes. The University of California/Caltech Keck Telescope (10 m), the Carnegie/Johns Hopkins/Arizona 8 m, the Arizona/Chicago/Italian/Ohio Columbus project (2 × 8 m), and the ESO VLT (15 m:4 × 7.5 m) are all likely to be operational in the mid 1990s, with 8–15 m National telescope(s) following in the late 1990s. In addition to these new-generation large telescopes, we will see the completion of several 2.6–4 m class instruments. These telescopes will be instrumented with the latest detector and instrumental technology, as will be current generation of 3–5 m class telescopes.

Will HST, a 2.4m telescope, with its 10–15-year-old technology, have a role to play? If so, will it play a major role, i.e., will it be at the forefront of scientific discoveries?

The answer is yes, but, with the qualification that the actual in-orbit performance of HST cannot, and must not, degrade significantly below nominal as-designed performance. If HST performs as initially planned in the areas noted below, it will be a major scientific success. If not, its scientific productivity and cost-effectiveness compared to ground-based facilities will have a major impact upon our ability as a scientific community to get the support we need for the next generation of large and immensely powerful space observatories.

The areas where HST has overwhelming advantages, in principle, over ground-based telescopes are:

(a) UV imaging and spectroscopy. This is a region of particular importance since it is inaccessible from the ground. HST, with its much larger aperture, will offer a major gain in capability over the very successful IUE telescope.

(b) Near-IR from $\approx 0.8\,\mu m$ to $\approx 2.5\,\mu m$. While not covered by any of the current, i.e., first-generation, instruments this wavelength region is one where the gains in space for background-limited observations are huge. The background in space between $1\,\mu m$ and $2\,\mu m$ is lower by at least two orders-of-magnitude, reaching *three* orders-of-magnitude lower at $1.6\,\mu m$. A high-throughput spectroscopic (resolution $R \approx 10^2$–10^4) and imaging system, as envisaged in the second-generation instrument program, will greatly enhance the scientific worth of HST. In fact, for the study of the galaxies in formation at redshifts $z > 1$, this region will prove to be of critical importance.

(c) High-resolution imaging and spectroscopy. Routine gains of an order-of-magnitude in the reduction in the FWHM of the point spread function (PSF; 'seeing'), from ≈ 0.5–2.0 arcsec with ground-based telescopes, to 0.03–0.07 arcsec with HST, will lead to dramatic discoveries. Seeing improvements of only a factor two at ground-based observations have led to very significant advances, particularly in the study of galactic nuclei and quasar 'fuzz'. There is no doubt that HST's much greater gain in resolving power will lead to major changes in our physical understanding in a variety of areas, particularly the environs of quasars, active and 'normal' galactic nuclei, distant galaxies, star clusters, and even star-forming regions, although, for this latter area improved IR capability would be of particular utility. The relative performance of HST and ground-based telescopes is demonstrated in Fig. 16.1.

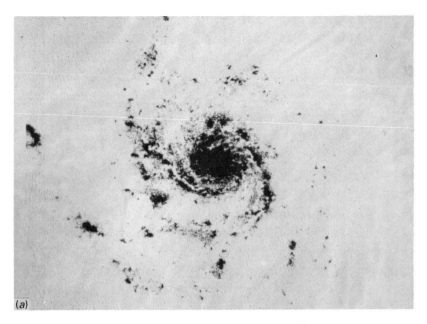

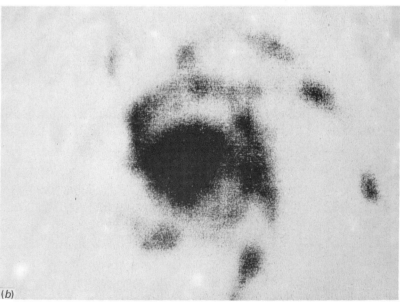

Fig. 16.1.(*a, b*) A simulation comparing the resolving power of HST with that of a ground-based telescope at typical, i.e., ≈ 1 arsec seeing. The galaxy is an Sc1, at moderate redshift ($z \approx 0.05$) as it would appear imaged both with HST and from the ground.

(d) Stable and calibratable PSF (point spread function) and throughput. The ability to carry out long-term programs in the optical and IR with a stable, well-calibrated and well-understood instrument/telescope system, without the throughput variations imposed by the atmosphere (on all timescales from scintillations with $\Delta T < 1$ sec to diurnal), is a capability unique to space-based facilities.

Clearly HST is a mission of outstanding scientific potential. Provided it is flown with its performance as originally conceived, with a viable plan for maintaining its critical limited-life components, and with instrument upgrades such as those under development in the ORI program (Orbital Replacement Instruments, i.e., two-dimensional spectrometers and near-IR imagers and spectrometers) HST is a mission guaranteed to be a great scientific success. Its unique capabilities will only be enhanced by the complementary strengths offered by the new large ground-based facilities. The spectroscopic power of these 8–15 m telescopes will dramatically improve our prospects for physically understanding and appreciating the implications of the discoveries made by HST.

2 The spacecraft – an overview

The Hubble Space Telescope is a very complex spacecraft, built to operate at a level of precision unique for astronomical missions. In particular, the figuring of the diffraction-limited optics, and the few milliarcsec precision required of the pointing control system, has pushed the limits of available technology. In addition, very severe demands are placed upon any high-technology development by the requirement that the instruments and supporting hardware must function for a period in excess of a decade in the harsh environment of space, with minimal opportunity for maintenance.

Yet, a telescope is a telescope. So some considerable understanding of HST can be obtained by breaking it down into its primary components, and identifying the role and expected performance of each. In fact, HST can be viewed as three readily identifiable subunits, the Optical Telescope Assembly (OTA; the 'telescope' itself), the Support Systems Module (SSM; the 'power and control systems'), and the Science Instruments (SIs).

The heart of HST is the OTA, the optical telescope assembly, identified in Figs. 16.2 and 16.3. This is a graphite-epoxy structure that supports the

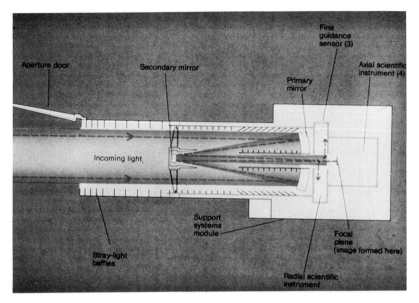

Fig. 16.2. A schematic of HST showing the light path. The three major components, the OTA, the SSM, and the SIs, can be identified. The outer shell is the SSM, the Support System Module, while the inner assembly that includes the primary, the secondary and their supporting structure is the OTA, the Optical Telescope Assembly. The SIs or Science Instruments, consist of the radial instrument, four axial instruments and the three FGS's, the Fine Guidance System.

primary and secondary, the interior baffle system, the instruments, and the critical components of the pointing control system (PCS), namely the fine guidance sensors (FGS), the star trackers and the rate gyros. The OTA, including the mirrors, was fabricated by Perkin-Elmer.

Surrounding the OTA is the SSM, the Support Systems Module. The SSM is constructed from several major modules as shown in Fig. 16.3. Together these form the outer structural shell, and act as a light shield, as well as playing a critical role in the thermal control of the telescope and the instruments through passive (insulation) and active (heaters) means. HST operates at 'room' temperature, i.e., the design operating temperature of the OTA is 18 °C. The SSM also provides additional baffling for the light path, supports the aperture/safety door, the solar panels and the communications antennae, and provides the supporting equipment for the operation of HST, namely the power system (batteries; charging, control and distribution hardware), the on-board computers and control system, magnetic torquers and reaction wheels (for pointing and tracking), and the communication

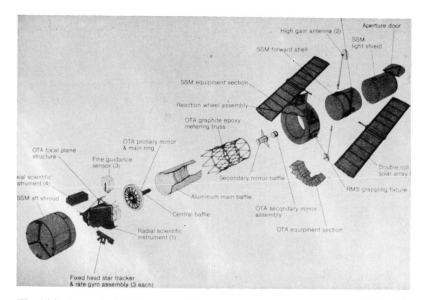

Fig. 16.3. An exploded drawing of the major support components of HST. All those to the left of, and including the secondary mirror assembly, are attached to or part of the OTA, with the exception of the Aft Shroud, which is part of the SSM, as noted. The remaining components to the right comprise the SSM.

system. Most of this hardware is located in the equipment section that surrounds the primary mirror.

Attached to the OTA are the items of particular interest to astronomers, i.e., the instruments. The instrument complement initially comprises two cameras, the CCD-based Wide-Field Planetary Camera, the WF/PC, pronounced *wiff pick*, and ESA photon-counting Faint Object Camera, the FOC, plus two spectrographs, both of which use Digicon detectors, the dual-channel (red and blue/UV) Faint Object Spectrograph, the FOS, and the UV only High Resolution Spectrograph, the HRS, and an image dissector scanner-based photometer system, the High Speed Photometer, the HSP. The FOC also has a limited long-slit spectrographic capability. In addition, the three fine guidance sensors, the Koester's prism interferometers that are used for precision guiding and offsetting of the telescope, can be used as instruments in their own right for astrometry. To minimize mechanical motions and the number of optical elements, all the instruments are fixed relative to the focal plane, as shown in Fig. 16.4. The characteristics of the instruments are summarized, rather broadly, in Table 16.1. Reference should be made to the STScI *Instrument Handbooks* for greater detail.

HST has been described in considerable detail in a variety of articles, as

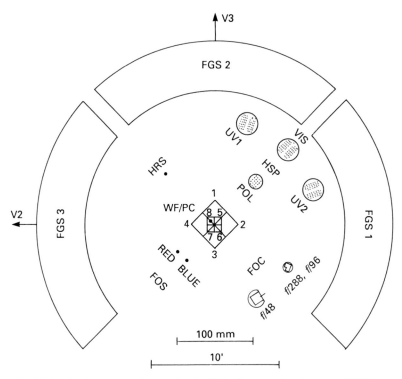

Fig. 16.4. The instrument layout in the 30 ft diameter focal plane of HST.

have the instruments. Further information can be found in Bahcall and O'Dell (1979), Leckrone (1980), O'Dell (1981), and Bahcall and Spitzer (1982), as well as in the Patras IAU meeting on 'The Space Telescope Observatory' (Hall 1982), and references therein, and the 1985 *STScI Call for Proposals for HST*, and the companion *Instrument Handbooks*. Following a summary of the expected performance of the telescope, and a short description of the PCS, the pointing and control system, some of the highlights of the test program will be given.

3 The spacecraft – expected optical performance

HST is an $f/24$ diffraction-limited 2.4 m Cassegrain telescope. The mirrors are ULE with MgF_2-overcoated aluminum, giving response from $\lambda \approx 1150$ Å to $\lambda \approx 1$ mm. The combined primary + secondary throughput is expected to be $> 60\%$ for $\lambda \geqslant 3000$Å, and $> 40\%$ for $\lambda \geqslant 1150$Å. Since the optical system is of Ritchey–Chretien design, the field is large and almost

Table 16.1 *Scientific instruments*

Instrument	Field	Resolution	Band (Å)	Limit
Wide field/ planetary camera	2.7';1.2' □	0.1";0.04"	1150–11 000 Å	~28[m]
Faint object camera	11";22" □	0.02";0.04"	1200–6000	28[m]
Faint object spectrograph	0.1" to 4.3"	3Å;30Å	1150–7000	22;26[m]
High resolution spectrograph	0.25" to 2"	0.3;0.15;1.5Å	1100–3200	11;14;17[m]
High speed photometer	0.4",1",10"	16μsec	1200–8000	24[m]
Fine guidance system	69' □	0".003	4670–7000	17[m]

*IDTs/PIs: Westphal; Macchetto; Harms; Brandt; Bless; Jeffreys

coma-free, being some 30 arcmin in diameter, but with significant astigmatism outside the central few arcmin. Optical correction for astigmatism is made within the off-axis instruments where necessary, as in, e.g., the FOC. The scale in the focal plane is 28 μm per 0.1 arcsec.

The design specifications call for 70% of the encircled energy within a radius of 0.1 arcsec at 6328 Å. The measured wavefront error is less than 1/60 of a wave rms at this wavelength. It is expected that this will result in images that meet the above specification, and which will have FWHM of 55 milliarcsec at 6300 Å, decreasing as expected for a diffraction-limited system to $\approx$ 30 milliarcsec at 3000 Å. The images are expected to have still smaller cores in the UV, but with wings that will become progressively broader as the wavelength decreases. These broad wings will ultimately contain most of the energy. They result from microroughness on the mirrors.

4 The spacecraft – pointing and tracking

The pointing and control system, the PCS, plays a critical role. To exploit the diffraction-limited performance of the optical system, tracking and pointing control must be better than 10 milliarcsec, as must the 'jitter' introduced by noise and resonances. In fact, the desired performance goal is $\leqslant$ 7 milliarcsec rms; some 1–2 *orders*-of-magnitude better than ground-based telescopes. Many aspects of the spacecraft need to work together to ensure that this goal can be met, and that the desired absolute and relative pointing capability is attained.

All maneuvering required for target acquisition and pointing control is carried out with four electrically driven reaction wheels located in the equipment section of the SSM; for reasons of redundancy there is one more wheel than the three strictly required. The wheels are used both for precision offsetting and tracking at the few milliarcsec level (while resisting torque from gravity gradient and atmospheric drag), and for slewing, thereby requiring operation ($\pm$) over a 175:1 torque range. The reaction wheels must be precision, low friction assemblies, particularly so since transitions through zero rpm may be required while tracking. Glitches must be minimized so as to avoid loss-of-lock by the FGS's. The maximum slew rate is limited to $\approx$ 6 deg *per minute*, the rate at which the minute hand of a watch moves. Clearly, slewing must be minimized.

Four magnetic torquers, that interact with the geomagnetic field, are also used for unloading the reaction wheels and for orientating the telescope is slew mode.

Coarse position sensing is done using four coarse sun sensors, and two 3-axis magnetometers that use a 36 parameter multiple-dipole model of the geomagnetic field; together these provide for absolute attitude determination to within 6 deg. The absolute attitude can then be established more precisely using the three Fixed-Head Star Trackers, the FHSTs. The three FHSTs operate independently of the main optics and can establish absolute pointings to ≈ 30 arcsec using bright $(2 \leqslant V \leqslant 5.7$ mag) isolated stars.

The primary reference for spacecraft orientation is a set of six (again redundant; four normally being used) dual-rate rate gyros, the RGAs, that are located on the OTA equipment shelf. These gyros are characterized by low noise and low drift rates; rms drift rates are 12 milliarcsec per second in the low rate mode. The RGAs are used in their high-rate mode, calibrated by the FHSTs, when necessary, for maneuvering. Short observations can be taken open-loop in low-rate mode, relying upon the RGAs to control the pointing. This may be necessary if, e.g., no guide stars can be found. Following calibration by the FHSTs, the drift rates could be as low as 6 milliarcsec per second. However, even with these low gyro drift rates, significant degradation will occur for observations requiring the highest resolution. For example, a 10 sec integration will be subject to ≈ 80 milliarcsec of image motion. The use of an FHST plus one FGS may provide a solution in some cases where only one good guide star can be found. While the pointing precision of the FHSTs is poor, their lever arm from the optical axis is large.

Normally, however, observations will be taken in closed-loop mode with two of the three fine guidance sensors 'locked onto' guide stars. In this mode, error signals derived from the FGSs are used to update the RGA-derived positions each second of time, giving pointing to 7 milliarcsec rms or better (the 'jitter' specification referred to above).

The FGSs are the heart of the fine pointing and offsetting capabilities of HST. Star selection from the 69 arcmin^2 field of each FGS is performed using a system of rotating mirrors, prisms, and refracting elements to direct the collimated wavefront from a stellar image to two crossed interferometers. Guide stars are acquired within a 5 arcsec 'aperture' set by a field stop, through a spiral search pattern. The interferometric systems then determine stellar positions to a precision of a few milliarcsec within their 'lock-on' range of ± 15–20 milliarcsec, and provide guiding signals for isolated stars brighter than $V \approx 14$–15 mag. Even fainter stars (possibly to $V \approx 16$–17 mag) can be measured for astrometric purposes using the remaining FGS, to derive relative positions with accuracies of 2–3 milliarcsec. Since the error signals from the Koester's prism interferometers are severely degraded when the guide stars are double with separations greater than ≈ 20 milliarcsec, several pairs must be uploaded to guarantee that a usable pair can be found. In fact, recent studies of binary frequency, by McAlister *et al.* (1987) and Shara *et al.* (1987), for this range of separations and the FGS magnitude range (≈ 8–15) indicate that three separate lists of up to six stars each may need to be uploaded and each possible pair acquired to ensure a high probability of success. Since acquisition and verification that the first pair is usable may take 5–10 minutes, and 1–2 minutes for subsequent pairs, the potential exists for significant loss of observing efficiency due to this problem.

The FGSs play a critical role for tracking moving targets, e.g., planets, features on planets or satellites of planets. The desired, often complex, track of HST can be carried out, in principle, by appropriately commanding the optical system used for stellar acquisition in the FGSs. The resulting error signals from the guide stars will ensure that HST follows the desired path. Executing such motions fully exercises the capabilities of the flight hardware and the ground system; further testing and development is needed, particularly in the ground system, to ensure that the moving target tracking capability becomes one of the routine operational procedures, albeit one of the most complex and demanding of all the procedures.

Ensuring that the pointing and tracking performance of HST is as desired has proven to be a complex task. It is complicated by the fact that many hardware and software systems play a role in this area. To establish that the desired performance could be reached, extensive testing of HST as a system was required, and this is, of course, a much more difficult, time-consuming and expensive process than testing individual components. Yet clearly this 'system-level' testing is essential before the spacecraft could be launched. The next section summarizes the activity that has occurred in this area.

5 The spacecraft–the A&V program

Lockheed Missiles and Space Company, LMSC, is the prime contractor for HST integration and has had the responsibility for bringing together the components, for assembling the spacecraft, and for testing the assembled vehicle. This program, known as the Assembly and Verification, or A&V, program ran for over a year and a half at LMSC, and culminated in

Fig. 16.5. The clean-room Vertical Assembly and Test Area (VATA) at LMSC. HST was assembled from its components in the vertical assembly stand. The OTA (standing vertically at left), four of the five instruments, the HSP, HRS, FOS and the WF/PC, in the lower centre, and the Equipment Section of the SSM in the upper centre.

the thermal vacuum/thermal balance test May/June 1986.

Components that had been tested by the subcontractor to ensure that they met the specifications, were brought to LMSC, where they were linked with their associated components, tested again, and so on as the spacecraft was assembled. Since the OTA is the core of HST, assembly occurred around this unit in a clean room of gigantic proportions. Great care was, and continues to be taken, to ensure that dust or chemical contaminants did not degrade the optical performance, particularly the UV throughput. Figure 16.5 shows the OTA and several of the instruments in the VATA, the Vertical Assembly and Test Area, at LMSC. Three of the four major components of the SSM are shown in Fig. 16.6, while the completed system is shown in Fig. 16.7 (devoid, however, of the highly reflective MLI, or Multi-Layer Insulation that now covers the upper sections). Fig. 16.8 shows the telescope being moved to a horizontal position in the VATA. It is in this completed state that several milestone tests were carried out to ensure that HST did perform as a mechanical/electrical/electronic/thermal/optical system. The five tests were

Fig. 16.6. Three of the four major components of the SSM. Left to right they are the Equipment Section that surrounds the primary, the Forward Shell which adjoins the Equipment Section, and the Light Shield, the uppermost of the SSM components which lies forward of the secondary and holds the aperture door.

the 'Functional', the 'Jitter', the 'Modal', the 'Acoustic', and the Thermal Vacuum/Thermal Balance (TV/TB) tests.

The first of these, the functional testing, was begun in the spring of 1985 as the spacecraft began its integration. These tests were essentially box-by-box tests of the entire electro-mechanical system. They were executed by an automated testing program, and grew in complexity as the spacecraft matured. After a baseline level of performance was reached, both abbreviated and complete systems tests preceeded, and followed, each of the environmental tests.

The second of these was the 'Jitter' test held late in 1985. During this test, the motion of critical components was monitored while possible sources of mechanical excitation that may be operated during observing were exercised. These include the reaction wheels, the FGSs, the tape recorders, and the filter or grating wheels in the SIs. To perform the test, the telescope was isolated by suspending it from air bags and the vibration levels measured with accelerometers. The actual measured value of jitter was found

Fig. 16.7. The spacecraft, complete except for its MLI, the Multi-Layer
Insulation, Equipment Section doors, and, of course, its solar panels and
communications antennae. The Aft Shroud, the fourth of the SSM components
is at the bottom here, below the Equipment Section.

to be close to that predicted and less than the required level of 7 milliarcsec;
the mechanically-induced line-of-sight jitter was found to be less than 5
milliarcsec.

Closely following this test was the 'Model' test, where the telescope was
shaken to identify the natural modes of vibration. Again the measured values
were close to those predicted. The lowest frequency mode was found to be
13 Hz. This is a scissor-like oscillation between the OTA and the SSM. HST
will, however, have much lower natural frequencies in its final flight

Fig. 16.8. The completed telescope being lowered to a horizontal position in the VATA.

configuration when the aperture door, antennas, and the solar panels are added. These frequencies will then be near 0.1–1.0 Hz.

In the 'Acoustic' test, held early in 1986, the spacecraft was exposed, in the acoustic chamber at LMSC, to sound levels comparable to those which it would receive in the shuttle bay during launch. This test was carried out to ensure that no damage occurred through acoustically-excited resonances. Subsequent tests verified that all systems functioned correctly. No serious problems were found, i.e., nothing major broke.

The most significant test of the A & V period occurred May and June 1986. This was the thermal vacuum/thermal balance test, where the fully assembled 43-ft long HST was placed horizontally in a large vacuum chamber at LMSC, and subjected to 58 days of 24 hr per day testing and operation. This test is the most realistic of the A&V process, being a system test of the spacecraft in a space-like vacuum and thermal environment, with the goal of verifying the thermal and power design, while exercising the scientific instruments and the support equipment.

The chamber was evacuated to $< 10^{-6}$ torr, and the satellite was subjected to realistic levels of solar irradiation, Earth back-warming, and deep-space cooling. The thermal performance of the spacecraft under extreme conditions, the 'hot' orbit and 'cold' orbit conditions, was also evaluated. All the instruments were operated in simulated observational modes, with data being collected not only to evaluate the performance of the instruments, but also to ensure that the data transmission and its transfer through the ground system occurred correctly.

This was also the time to verify that *photons in* at the primary resulted in *electrons out* from the detectors and *transmitted bits* from the electronics. Thus, end-to-end optical throughput tests of the whole system were also performed. A particular motivation for this was the concern that molecular contamination during assembly may have degraded the UV throughput of the system. In addition, the optical-throughput test would allow first-order cross-calibration between the instruments, and ensure that there were no major obstructions or vignetting of the light paths, or ghosts or light leaks. These tests were carried out using a set of NBS-calibrated lamps, and generally verified the throughput predicted from the individual instrument calibrations. There was no evidence of molecular contamination.

Furthermore, a test in which scattered light was measured by one of the WF/PC CCD's was used to show that the $< 20\mu$m dust accumulation onto the primary and secondary mirrors was within acceptable limits, i.e., the primary obscuration from particulates is $\approx 3\%$, when both the photographic $> 20\mu$m measures and these results are combined. The obscuration on the secondary is significantly less. The tests also verified that the secondary mirror would lie well within its focus range when the optical system was focused. This was very reassuring.

In general, the A&V program, including the TV/TB test, was very successful. The areas of concern are discussed later. Some of the results from the A&V program, and the TV/TB test, are summarized here:

(a) *UV/optical throughput* was found to be good over the range 1216 Å–7000 Å. There was no detectable molecular contamination. Dust accumulation and obscuration were within acceptable limit at $\approx 3\%$.

(b) *SI throughputs* were found to be consistent (at 20–30% level) with the *Instrument Handbook* estimates.

(c) *Jitter* was less than 7 milliarcsec rms, ensuring minimal image degradation.

(d) *Data transmission* and bit-error-rate performed within

specifications for each of the data rates, i.e. 4 and 32 Kbits s^{-1}, and 1 Mbit s^{-1}.

(e) *SI Parallel operation*, namely that of the WF/PC and all axial pairs was validated. In particular, WF/PC + FOC + one of HSP, FOS or HRS could operate together (but see below).

(f) *System compatability tests*, structured to demonstrate that the subsystems would work together as required for routine in-orbit operation, without interfering with the science data, were very successful.

(g) *SIs*: the results from the instruments were mixed, but generally positive. The value of operating HST *as a system* in a realistic environment was clearly demonstrated by the detection of several problems that can now, in most cases, be fixed before launch. The major problems detected ranged from a serious light leak in the HRS from off-axis light passing through a hole in the aperture plate, to noise spikes and intermittent power supplies in the HSP, to low UV-throughput in the WF/PC, apparently from icing. The 1986/87 STScI newsletters discuss these and other instrumental problems, and the fixes that are being implemented.

At the end of TV/TB an important end-to-end operational test of HST was carried out—the Ground System Thermal Vacuum test, or GSTV test. HST was operated as though it was in orbit. A science mission specification (SMS) was generated at STScI from six proposals entered into the HST scheduling system, the science planning and scheduling system (SPSS), and sent to the control center at Goddard Space Flight Center across our high speed connecting link, where they were used to construct the spacecraft command loads for the test. The command loads were transmitted to HST and executed. For 3.5 hours HST 'performed' a series of observations with the WF/PC, FOS and HSP. The test involved two spacecraft 'slews', wrote data to a tape recorder, and read out the various instruments ten times. The expected data were received and monitored by the Observation Support System (OSS) at Goddard and STScI. The command loads executed flawlessly. The test was a milestone in that it represented the first observations actually executed by the actual planning and commanding systems that are to be used during normal HST operations.

Further end-to-end tests are planned during 1987. At those times it is expected that the throughput of the ground system will also be evaluated.

While the first GSTV test executed flawlessly, it did so only after a Herculean effort on the part of the ground system operations and software staff. The efficiency was between *one and two* orders-of-magnitude worse than that which would be acceptable in orbit. Schedules and commands, clearly, must be generated in better than real-time if the overall efficiency of operation of HST is not to be impacted.

The spacecraft survived the rigors of the LMSC test program and performed well. However, there were two areas, thermal and power, where the spacecraft's performance was not as expected. First, the spacecraft's thermal performance differed from that predicted by the thermal models–it ran too cold, particularly the primary and secondary mirrors, which were found to be operating some 3° C cooler than expected, even with all heaters on. Control was established, but with reduced margins. The OTA needs to operate near its design temperature and within heater control range to ensure that the optical performance goals, namely pointing stability and wavefront error are met. In addition, the need to operate more heaters than expected contributed to the greater-than-expected power consumption. Rework of the OTA/SSM thermostats and insulation is expected to lessen the thermal problems, probably to the point where they will be of little concern, except for the impact of the higher heater power usage upon the power situation. It is, however, in this second area, that of power consumption, that HST faces a more serious problem.

Power consumption during TV was some 8% (200 W) higher than predicted. Since TV, a very careful analysis of the power requirements of HST has been carried out, with some disturbing results. The most significant has been the realization and acceptance that the NiCd batteries must be trickle-charged for a significant fraction of the sun-illuminated portion of the orbit–for possibly as long as 20 minutes, or about one-third of the available charging time. If they are not routinely trickle-charged, the battery lifetime is short and unpredictable.

As the system stands, the available time *initially* for trickle charging is the needed ≈ 20 minutes per orbit. This assumes the currently expected output from the solar panels (recently derated by 4% from a balloon recalibration of the cells), and the higher than expected power consumption. However, it does require that the power consumption of the FGSs and the SIs be minimized, by limiting the number powered-up for operation to, for example, two FGSs plus the WF/PC and one Axial SI. Unfortunately, during normal operation later in the mission (on a still uncertain timescale of 1–2 years), the available time for trickle charge would fall below the desired value as the solar arrays age. Other operational constraints would then be required on, for example, roll and the general sun-spacecraft attitude.

These operational 'fixes' will impose serious limitations on the science program, and on the efficiency of implementation of the science program. For example, the efficiency will be significantly lowered if the number of instruments that can be powered at any time is reduced, since all instruments require long warm-up periods (e.g., 12 hours for the FOC and the FGSs). If only two FGSs can be used, not only the efficiency but also the operational capability to actually carry out many programs will be impacted. Certain targets may be inaccessible due to no guide star pairs, or so few stars that the probability of a successful acquisition is too low. Planetary, astrometry, transient-event, and roll-dependent observational programs will be impacted.

Furthermore, such 'operational' fixes are the only available contingency once the spacecraft is in orbit. There is clearly a significant element of risk associated with limiting the operational options, and thus the contingency against further unexpected power problems once in orbit. At best the science mission could be severely constricted, while at worst the spacecraft itself may be imperiled.

Solutions that can be implemented on the spacecraft before launch are the most desirable. The obvious one is to minimize power consumption (through reduced heater load). Further examples would be to replace the batteries with some which do not require a period of trickle-charge, or at least, less trickle-charge, or to increase the output of the solar panels. The most desirable approach would be some combination of these. Such hardware solutions are still under consideration as a means to rectify the power problem, but it remains our foremost concern.

6 The ground system

Most of the discussion and publications to date regarding HST have concentrated on the spacecraft and on the instruments. From the point of view of the scientific returns to be derived from HST, the ground system deserves equal time. If the ground system does not work well, many science programs, as well as the overall efficiency, will be severely impacted. Yet the ground system is complex, because of (i) the complexity of HST, (ii) the difficulties of operating in low-earth orbit, and (iii) the range of tasks that it must perform. For example, the ground system encompasses all the links in the chain from proposals, to scheduling, to commanding, to real-time observations, to receipt of data, to preliminary processing and archiving, to monitoring, and to support of data analysis. To illustrate the complexity of developing a workable, and efficient, schedule and program, some examples of both the constraints on the system, and the capabilities needed to carry out a realistic science program with HST, are given:

(i) *Low Earth orbit.* At about 600 km, the orbital period is only 95 minutes, of which only ≈ 35 minutes are dark. A significant fraction of programs will require 'dark time', even though much effort has gone into minimizing scattered light.

(ii) *Very slow slew rate.* At 6 deg per minute, and a 95-minute orbit, slewing must be minimized so as not to seriously degrade efficiency.

(iii) *TDRSS access.* Limited access to TDRSS, the Tracking Data Relay Satellite System, requires that all observations be pre-planned. Real-time interaction for target acquisition and SI parameter adjustment is limited to $\approx 20\%$ of the time.

(iv) *Preplanned observations.* The need to preplan introduces, in itself, further problems; e.g., the large volume of information that must be received, entered, checked and verified before scheduling can be done.

(v) *Data volume limits.* TDRSS is needed for 1 Mbit s^{-1} data transmission. On-board tape recorders will handle data when TDRSS is unavailable, but planning is required to ensure they do not fill up before a TDRSS access occurs.

(vi) *Pointing constraints.* There are many, most of which are soft, i.e., they can be violated under certain circumstances, thereby adding to the scheduling complexity. They are: 50 deg from the Sun; 30 deg from the full moon; 80 deg from the bright Earth limb; 5 deg from the dark Earth limb; orientation with respect to the ram direction–certain observations may require the telescope to be pointed away from the direction of motion, to minimize contamination by the spacecraft 'glow', i.e., light from collisional excitation of atmospheric molecules. Even the Sun constraint is somewhat soft. To observe Mercury for which $\Delta\theta < 50$ deg, the telescope must be slewed to the planet during dark, the planet observed as it 'rises' across the Earth's limb, and the telescope slewed away before the Sun 'rises'.

(vii) *SAA passage.* The spacecraft passes through a small portion of the South Atlantic Anomaly for $\approx \frac{1}{3}$ of its orbits. The high radiation level will preclude observations during passage, *and* will limit the type of observations that can be done following passage due to high radiation-induced backgrounds in the instruments. The use of the instruments is further complicated following passage, since the decay rate of the background is different for each instrument.

(viii) *Solar panel alignment.* The need to ensure adequate power output from the solar arrays adds roll constraints. Deviations in roll of ± 5 deg are always allowed, with roll deviations of ± 30 deg allowed for $< 10\%$ of the time. The power problem discussed above may preclude *any* roll beyond the nominal ± 5 deg.

(ix) *Guide star acquisition.* The FGS's can take up to 10 minutes to acquire guide stars. Since not all pair acquisitions will be successful (e.g., some will be binaries), additional time, at ≈ 1–2 minute per pair, will have to be allocated to allow for further attempts at finding a good pair. This time cannot be recovered even if the first acquisition is successful. HST runs on an 'absolute' clock.

(x) *Guide star availability.* Suitable guide stars are not always available at the time chosen by the scheduling program for the observation; the calender has to be iterated. When this happens often, as it appears will be the case, the time and effort to converge to a practical calender become large.

(xi) *Uncertain orbital phase.* The phase of HST in its orbit cannot be precisely established due to unpredictable atmospheric drag. HST's position in its orbit 1.5 months in advance will likely be uncertain to 20%, i.e., up to 8000 km. This makes scheduling for time-critical observations rather difficult.

(xii) *Data volume.* HST will generate 0.4 *Gbyte* of data per day. This is comparable to the VLA data rate and to the *mean daily sum* of all digital ground-based optical data (1984 numbers; it is clearly more now). These data must be processed, catalogued, archived, and prepared for distribution to the scientist concerned as quickly as possible.

(xiii) *Instrument modes.* Between the five instruments and the FGS's there are some 5.10^4 modes (filters, gratings, detectors, etc) to be commanded and calibrated.

(xiv) *Calibrations.* Instrument and astronomical (i.e., absolute) calibration procedures must be devised, and the calibrations themselves proposed, scheduled, catalogued, processed, verified, archived, and made available to the pipeline calibration system and to the user who wants to verify the calibration personally. Absolute calibration data must be available through databases.

(xv) *Command validity.* The validity and uniqueness of the command loads must be ensured. While the spacecraft is not too fragile, the consequences of mistakes do not bear thinking about.

(xvi) *Proposal volume.* Some 1000–2000 proposals per year must be processed, peer-reviewed, and those several hundred that are accepted must be checked for technical problems and transferred into SPSS to be scheduled. Because of the preplanned observing process, each is significantly more complex than ground-based proposals.

(xvii) *Dealing with the high spatial resolution.* The astronomical community is not prepared to deal routinely with the difficulties of preplanned hands-off target acquisition at the < 0.5 arcsec level. Considerable effort will be required to ensure that HST time is not wasted because of astrometric uncertainties and incorrect aperture placement.

The task awaiting the ground system is clearly formidable. In fact, at STScI alone the various software systems total more than *one million lines-of-code*, where line-of-code means executable source code. This is nibbling at the lower fringes of SDI territory. The Science Operations Ground System, SOGS, forms the largest fraction of this. This system was developed by TRW at a total cost of some \$65 M, including ≈ \$8 M for hardware, and consists of three major components: SPSS, the Science Planning and Scheduling System; OSS, the Observatory Support System; and PODPS, the Post-Observation Data Processing System. In the simplest of terms, SPSS takes proposals and does calender generation, scheduling, incorporation of guide star positions, and generates a set of commands (the Science Mission Specification or SMS) that is transferred to the HST Control Center at Goddard for further enhancement, resulting in the full command load that is sent to the spacecraft. OSS's role is to support the real-time observations, while RODPS involves data receipt, pipeline calibration and preliminary archiving.

To fully develop the roles of these and the other STScI and Goddard software systems would result in a review of considerable size. I refer the reader to the 1985–1987 STScI newsletters as one source of information on these systems.

Conclusions

I have emphasized the complexity of the HST program, from the flight hardware to the ground support. Yes, it is complex – but there is no fundamental technical reason that HST should not fulfill its great potential to dramatically expand our knowledge of the universe. Success, as always, is a function of the support and commitment of those involved in the mission. In an era of major advances in ground-based capability, HST still stands out for its uniquely powerful capabilities. In fact, the complementary capabilities offered by these new large ground-based telescopes will greatly enhance the scientific value of the discoveries made by HST.

Acknowledgements

Many scientists and engineers (both hardware and software) have contributed to my understanding of HST. I am grateful to them for their time and their help. In particular, I would like to thank Pierre Bely, Roger Doxsey, and Peter Stockman for their careful reading of this manuscript, and for their many comments. However, any errors of fact, as well as opinions expressed, should be laid at my door, and not upon them.

References

Bahcall, J.N. & O'Dell, C.R. 1979, 'The space telescope observatory', In: *Scientific Research with the Space Telescope*, eds M.S. Longair & J.W. Warner. NASA CP-2111.

Leckrone, D.S. 1980, *PASP*, **92**, 5.

O'Dell, C.R. 1981, 'The space telescope', in: *Telescopes for the 1980's*, eds G. Burbidge & A. Hewitt. Palo Alto, California: Annual Reviews, Inc., p. 129.

Bahcall, J. & Spitzer, L. 1982, 'The space telescope', *Sci. Am.*, **247**, p. 40.

Hall, D.N.B., ed., 1982, *The Space Telescope Observatory*. Baltimore: Space Telescope Science Institute.

McAlister, H.A., Hartkopf, W.I., Hutter, D.J., Shara, M.M. & Franz, O.G. 1987, *A.J.*, January.

Shara, M.M., Doxsey, R., Wells, E. & McAlister, H.A. 1987, *PASP*, March.

Note added in proof:

The major concerns of 1986/1987 expressed here have been overcome in the past year. The power situation is greatly improved through the development of improved solar panels and batteries. The projected operational efficiency of the SOGS/SPSS system is now likely to be close to the desired goal. While the operation of HST will be a tough and challenging task, I expect that the scientific returns will be as large as we had hoped, and I am looking forward to an exciting time when HST is launched in late 1989 or early 1990.

Garth Illingworth

17

The low dispersion survey spectrograph

KEITH TAYLOR

Summary

The Low Dispersion Survey Spectrograph (LDSS) is a wide-field, broad-band (350 nm–700 nm) grism spectrograph designed principally for multi-slit spectroscopy of faint sources at reciprocal dispersions between 160 and 1000 Å mm^{-1}.

1 History

The primary motivation for the project came from the realisation that, with the advent of very large optical telescopes (8 m and above), together with the Hubble Space Telescope, spectra of objects, significantly fainter than the sky-background, will be studied routinely. However, the plethora of relatively normal galactic stars and distant, but not necessarily interesting, galaxies to be found at these faint magnitude limits requires that some form of faint spectroscopic survey be attempted.

Fibre spectroscopy, as pioneered at the AAT, demonstrated the dramatic gains that could be made over conventional techniques for survey work. However there was a clear need for similar multi-object gains to be made at magnitudes significantly fainter than the sky-background, and hence beyond the normal scope of fibre techniques. The excellent throughput of Wynne's Faint-Object Spectrographs set a bench mark for all low dispersion systems. However, we were unable to accept its restriction in dispersion and, furthermore, required a significantly larger field in order to do multi-object work satisfactorily.

A consortium of motivated participants, consisting of R.S. Ellis

(Durham), P. D. Atherton (Groningen) and myself were thus formed around the Low Dispersion Survey Spectrograph concept. We were fortunate to have working for us C. G. Wynne and S. P. Worswick (RGO) who were stimulated to produce a revolutionary new optical design for us and R. N. Hook (UCL) who produced the software.

2 Design considerations

In designing the instrument we accommodated the following requirements.

(i) The system is optimized for multi-object survey work using multi-slits in order to facilitate precision sky-subtraction for sources considerably fainter than the sky background.

(ii) The instrument has a choice of dispersions since, for faint sources, the optimum spectral resolution is a strong function of the type of survey work to be undertaken.

(iii) It has a substantial blue throughput between 350 and 500 nm (with the appropriate detector) rather than being optimised for work longward of 700 nm where sky subtraction difficulties severely limit the types of objects that can be studied.

(iv) It has an externally accessible focus, allowing a variety of detector systems, both photon-counting and CCDs, to be used.

(v) Everything was done to optimize the overall throughput of the system.

These considerations led the optical design team to a fully transmitting collimator/camera utilising FK54, a new dispersion Schott glass. Two grisms, giving dispersions of 165 and 850 A mm^{-1}, were chosen for the initial observations. A practical upper limit to the field size of such a system, mounted at the Cassegrain focus of a 4 m telescope, is ~ 12 arcmin (Wynne & Worswick, 1986). For an $f/2$ camera this matches, approximately, the format not only of the IPCS but also the large-format CCDs under development and 1 (or 2) $\times$ 2 arrays of standard-format CCDs.

The focal reducer design allows great flexibility in the way in which the spectrograph can be operated. Four main modes of operation are allowed. (i) Direct imaging (no grism or aperture plates); (ii) Long slit (grism and long silt mask); (iii) Slitless mode (grism but no aperture plates); (iv) Multislit mode (grism with aperture plate mask).

3 Aperture mask

The process of design and manufacture for the multi-slit masks is a complex one involving fairly sophisticated software procedures coupled with a reliable manufacturing process which produces masks to the required accuracy. Ideally, one would like a mechanism whereby 2D images of a given patch of sky could be acquired by LDSS, in direct-imaging mode, and then immediately analyzed so that suitable masks could be manufactured on-line, at the telescope, for use the same night. However, we have yet to acquire the necessary facilities for on-line mask manufacture, and so we have organised our observations around the availability of AAT prime focus plates which allowed us the required accuracy in photometric and astrometric precision using the COSMOS and APM measuring machines.

With a data reduction software package which incorporated within it a

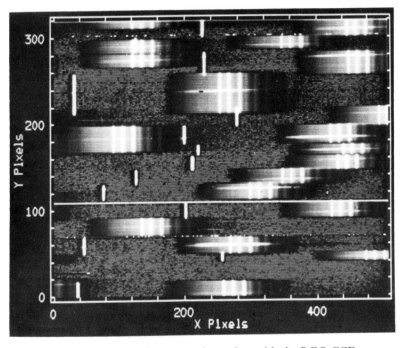

Fig. 17.1. An LDSS multi-slit observation, taken with the RGO CCD camera, at the AAT in May 1986. The exposure time is 4000 sec. For each slit, a zero order spectrum forms a bright vertical line, with the first order spectrum nearly half the frame to its right; the bright [OI] and NaI night sky lines are clearly visible. Each object is a horizontal band within the slit's sky spectrum. At 850 A mm^{-1}, spectra of detected objects in the B magnitude range 21 to 23.5 are clearly seen.

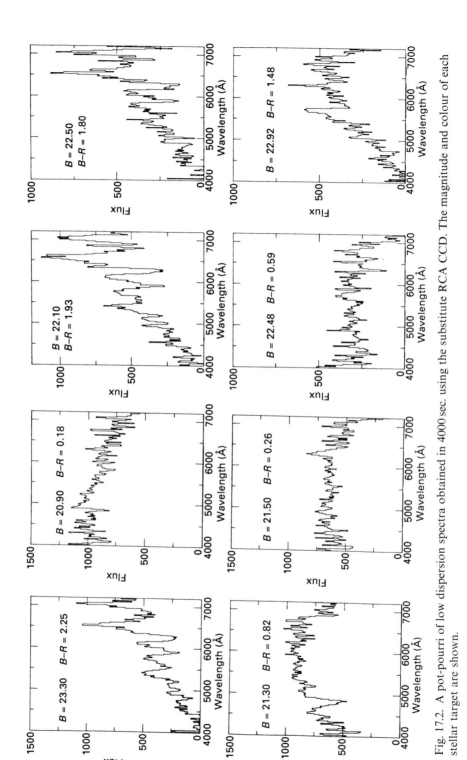

Fig. 17.2. A pot-pourri of low dispersion spectra obtained in 4000 sec. using the substitute RCA CCD. The magnitude and colour of each stellar target are shown.

detailed model of the optical and spectroscopic characteristics of LDSS, we are able to accurately predict the resultant image for any given multi-slit mask and detector configuration. This facilitated relatively simple off-line mask design, and allowed us to select a photo-chemical etching process for mask manufacture, which we found best suited our required accuracy.

We plan in the future to rationalise the mask making procedure in order to facilitate on-line manufacture. However, in order to match the accuracy and objectivity achieved at present, we will require very much more powerful computational facilities and software. The on-line manufacturing processes used at present would also have to be radically refined.

4 Commissioning results

The initial observations were made at the AAT in June 1986, using a low-grade RCA chip. Despite a high 105 electron rms readout-noise we were able to obtain multi-spectra images of point sources down to $B = 23.5$ mag

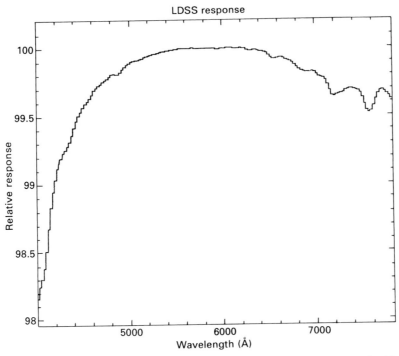

Fig. 17.3. Relative response of LDSS via observations of Feige 110 with low dispersion grism.

in ~1 hour, using the low dispersion grism. Fig. 17.1 shows a typical example. These images were subsequently reduced to sky-subtracted, wavelength and flux calibrated spectra using the LDSS software package on the RGO starlink VAX. Fig. 17.2 gives a selection of spectra all taken from the same 4000 sec image.

Using a flux calibration standard we were able to confirm our laboratory measurements which gave the LDSS optics a total throughput of ~60% at peak. The transmission with wavelength is shown in Fig. 17.3 where the peak has been normalized to 100%.

5 Acknowledgements

As always, this type of instrumental project could never see the light of day were it not for the efforts of many support engineers and technicians. Our thanks are particularly directed to Geoff Holt and Jim Lester (RGO), Mike Breare and John Webster (Durham University) and the support staff at the AAT.

18

Correction of atmospheric dispersion

C.G. WYNNE and S.P. WORSWICK

Summary

Atmospheric dispersion introduces serious limitations to observations with modern optical telescopes. A design for an atmospheric dispersion corrector is described which allows for the wide fields and wide spectral coverage demanded by astronomers using modern large telescopes of relatively short focal length.

During the 1860s the Astronomer Royal, Airy, was observing transits of Mercury and Venus, and he found accuracy of measurement seriously limited by atmospheric dispersion. Airy was interested in instruments, so he proceeded to invent a series of devices to correct this. His last and best attempt was an eyepiece modified so that the eye-lens could be tipped, thus effectively introducing a prism whose angle could be chosen to correct dispersion at the zenith distance of his object (Fig. 18.1). This worked admirably. It was, Airy (1869) wrote, 'a construction that for simplicity and optical perfection is not likely to be improved', and he was right. In 1869 he had solved the problem of correcting atmospheric dispersion.

It has not, of course, stayed solved, because the problem has changed. Visual observation has been replaced by detectors of higher sensitivity, recording observations over wider fields, larger spectral ranges, wider fields became available on larger telescopes, and recent devices have extended the range of simultaneous observations of many objects, e.g. multi-slit spectrographs, fibre feeds. The efficiency of these is seriously limited by atmospheric dispersion but, since Airy, remarkably little has been done to correct it. This paper described what we are doing to provide a cure, applicable to modern observing methods, and in particular for our telescopes on La Palma.

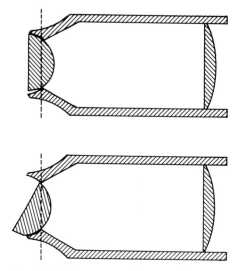

Fig. 18.1. Telescope eyepiece modified by Airy so that the eye-lens could be tipped to correct for atmospheric dispersion.

Dispersion has been corrected in special cases. A speckle camera built at Imperial College in 1976 (Beddoes *et al.*) employed an atmospheric dispersion corrector (ADC) consisting of a pair of zero-deviation prisms, independently rotatable so that their combined dispersion could be tuned to correct for the atmosphere at different zenith distances (Fig. 18.2). A similar camera built at Kitt Peak had the added refinement that the adjacent surfaces of the doublet prisms were oiled together to eliminate loss of light from two air-glass surfaces. These systems work well in collimated or near-collimated light, but in light with the levels of convergence of Cassegrain telescope foci (typically from $f/15$ to $f/8$), they introduce serious aberrations.

Mathematical analysis (Wynne & Worswick, 1986) shows that, for an ADC of this form for use at a Cassegrain focus, the asymmetric aberrations arising at the tipped prism surfaces can be made negligibly small by a correct choice of the glasses used. There then remains a single dominant aberration, a chromatic difference of focus arising simply from the thickness of the device. The minimum thickness is determined by the prism angles and the size of image field to be corrected, and this introduces a chromatic aberration of a size to make the simple system useless as a Cassegrain corrector. This aberration, however, can be simply overcome. If the doublet prisms are

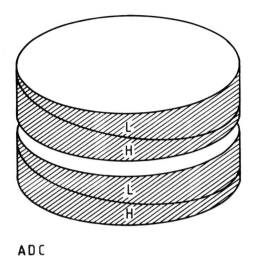

ADC

Fig. 18.2. An atmospheric dispersion corrector (ADC) consisting of a pair of independently-rotatable zero-deviation prisms.

arranged so that high and low dispersion glasses alternate through the stack, then the interface between the two prisms will separate glass of different dispersions and the aberration can be corrected by the substitution of a suitably curved surface interface for a plane one. This leads therefore to a simple ADC for a Cassegrain focus, having only two air-glass surfaces, producing negligibly small changes to the telescope imagery, apart from correcting atmospheric dispersion. A design for the $f/11$ focus of the 4.2 m William Herschel Telescope (WHT), correcting dispersion up to 70 deg zenith distance, covers a field of 15 arcmin diameter. The aberrations of the corrector, over the spectral range 365–1014 nm, give image spreads within $\frac{1}{10}$ arcsec, and at the extremes of the wavelength range and field diameter additional chromatic aberrations of up to $\frac{1}{4}$ arcsec. The field of good resolution of the telescope Cassegrain focus is of course much smaller than 15 arcmin. Correction of atmospheric dispersion at a Cassegrain focus turns out, therefore, to be quite simple.

Correction at prime focus presents a different problem. Types of aberration in the simple prismatic corrector, that are insignificant at relative apertures up to $f/8$, increase as quite high powers of the numerical aperture, becoming intolerably large at prime focal ratios of $f/3$ or faster. Our 4.2 m mirror works at $f/2.5$, and successive generations of large telescopes are

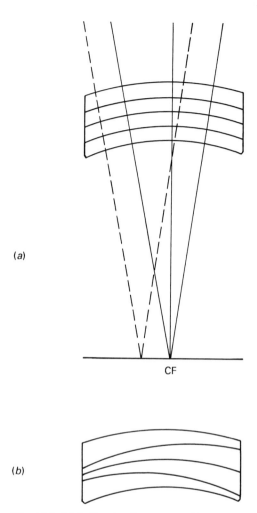

(a)

CF

(b)

Fig. 18.3. (a) Indicating how a set of lens surfaces all centred on the telescope focus is completely free of aberrations for axial imagery and nearly so for off-axis imagery; (b) a prime focus ADC consisting of two rotatable doublet zero-deviation prisms with spherical optical surfaces all centred on or near to the telescope focal plane.

getting faster, so a different form of ADC is required, in which these large aberrations do not arise.

Again, we have found a quite simple solution. It is obvious that a set of lens surfaces, all centred on the telescope focus, is completely free of aberrations for axial imagery–light from the telescope mirror at any aperture and all wavelengths is orthogonal to all the surfaces (Fig. 18.3a). Moreover, it follows from simple analysis that for off-axis imagery it is free from all first order aberrations of point imagery except for an astigmatic aberration–a tangential field curvature–and the system is also free from a number of higher order aberrations.

Such a system can provide the basis for a prime focus ADC, consisting again of two rotatable doublet zero-deviation prisms, but with spherical optical surfaces instead of plane ones, the surfaces being all centred on, or near to, the telescope focal plane (Fig. 18.3*b*). Again, mathematical analysis shows that the aberrations arising from the inclined surfaces can be made very small by choice of the glasses used, and the overall performance can be very high, apart from the inherent tangential field curvature.

This is an aberration causing an image spread increasing as the square of the field size and, in practical cases, it is not very large. For example, with a $f/3.2$ mirror, this ADC aberration would give image spreads within $\frac{1}{2}$ arcsec up to a field diameter of $\frac{1}{2}°$. This would increase with numerical aperture and field size, so for our $f/2.5$ mirror on a very good observing site it needs correcting–and this can be done.

The field of good resolution of an $f/2.5$ paraboloid used alone is very small indeed–less than $\frac{1}{2}$ arcmin for $\frac{1}{2}$ arcsec imagery. A prime focus corrector is therefore normally necessary, and an ADC would be used with it. A suitable location for the ADC would be immediately in front of the field corrector. It therefore seems reasonable to consider the two together, and it then emerges that the astigmatic aberration of the ADC can be readily removed by appropriate design of the field corrector. When this is done, the small residual aberrations of the combined system are effectively those of a field corrector designed to be used alone, the addition of an ADC (with only two air-glass surfaces) causing no further loss of image quality. This is better than might have been expected, and it provoked two further trains of thought.

In the first place, it made us look more closely at field correctors, which limit performance of the combined systems. It was originally intended that the prime focus of the William Herschel Telescope would be provided with a field corrector consisting of three separated thin lenses, of the general form used on the Isaac Newton Telescope, and the Canada–France–Hawaii Telescope. These work well up to apertures of $f/3$, but at $f/2.5$, over the wide field and spectral range that astronomers want, a triplet corrector does not do justice to the very good seeing that we now know to exist on La Palma. So we studied field corrector design at $f/2.5$ or faster.

The first lenticular field correctors covering fields of upwards of $\frac{1}{2}$ degree diameter (Wynne, 1967) designed for the Palomar 200 in and widely used,

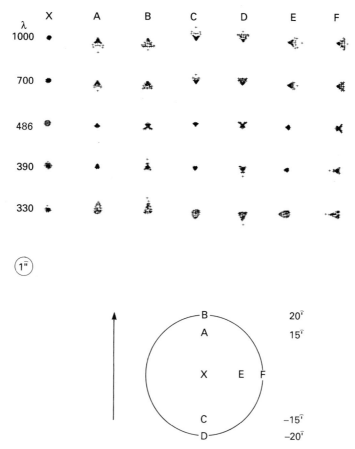

Fig. 18.4. Image quality obtained with a four lens prime focus field corrector behind a near-monochromatic ADC.

consisted of four separated lenses. These were later superceded (Wynne, 1974) by a triplet design, which performed as well as the quadruplets up to apertures of $f/3$, which was as fast as was then considered. But at $f/2.5$ or faster, we now find that a four-lens corrector can give substantially better imagery.

The best available combined field corrector system, to give a field of 40 arcmin diameter (unvignetted) on the William Herschel Telescope appears to be a four-lens field corrector, behind a near-monocentric ADC as described above. Figure 18.4 shows geometric aberrations for such a system, over the spectral range 330–1000 nm, with an ADC set to correct for a zenith distance of 70 deg (where its aberrations are maximum). Figure 18.5 shows the form of the device.

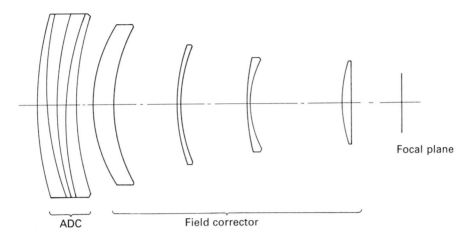

Focal plane

ADC   Field corrector

Fig. 18.5. The form of the prime focus combined corrector giving the
performance shown in Fig. 18.4.

The aberrations of Fig. 18.4, with a four lens corrector, still arise mainly
from the corrector, not the ADC. This suggests a further development. An
ADC could be designed with wider angle prisms, so that when tuned to
maximum, it would introduce much bigger dispersion than needed to correct
the atmosphere. Then, for atmospheric correction, the prisms would need
only small rotations from the null position. At wider angular settings the
device could be used for objective-prism type slitless spectroscopy. The
wider prism angles would require a somewhat thicker ADC, but the
consequent aberration could again be harmlessly removed in the field
corrector design. A slitless spectroscopy option could thus be added to the
telescope at very small extra cost.

Our ADC with prism angles nine times that required for atmospheric
correction on the WHT would give a stellar spectrum some 2.2 mm long,
comparable with that provided by the Southern Schmidt with its most
widely used objective prism, but with a plate scale 3.5 times larger, so that
crowding of spectra would be correspondingly reduced. Such an extra-
powerful ADC, used to correct the atmosphere, adds negligibly to the
corrector aberrations–Figs 18.4 and 18.5 relate to such a system. Tuned to
maximum dispersion, there is some increase in aberrations, but image
spreads remain within $\frac{1}{2}$ arcsec over most of the field and wavelength range.

The work described above has produced a range of options for field and
dispersion correction at prime focus. For smaller field sizes, or for observing
sites with average seeing, cheaper and simpler systems, with triple correctors

and smaller in size, may be adequate. For the highest image quality, over wider fields more advanced systems are needed, but they are still essentially simple.

References

Airy, G.B. (1869). *Mon. Not. R. Astr. Soc.*, **29**, 233.
Beddoes, D.R. *et al*, (1976). *J. Opt. Soc. Ann.*, **66**, 1247.
Wynne, C.G. & Worswick, P. (1986). *Mon. Not. R. Astr. Soc.*, **220**, 657.
Wynne, C.G. (1967). *App. Opt.*, **6**, 1227.
Wynne, C.G. (1974). *Mon. Not. R. Astr. Soc.*, **167**, 189.

19

Image photon counting

A. BOKSENBERG

Summary

The image photon counting technique has attained widespread use in astronomy. The principles are set out here, together with some recent astronomical results and some projected developments.

1 Introduction

I first thought of the image photon counting technique while soaking in the bath. In 1968, when I was at University College London, I had been made a member of a panel which had been set up in the UK to consider what instrumentation would be needed for the coming Anglo–Australian Telescope. My responsibility was to think of the detectors. I reviewed all the available devices and tried to anticipate where new developments would lead. However, the prospect of an efficient, common-user imaging detector device which did not degrade weak signals while having the capability also to record strong signals seemed very far away. Most detector devices produced commercially are developed for broadcast or military markets, and because of the great development effort required to make available any new device in this difficult field it had been advantageous to employ the more appropriate of these in astronomy, but they all had their limitations. Particularly, I was contemplating the possibilities for a spectroscopic detector. Spectroscopy is the most important investigative technique applied in astronomy; it accounts for the use of close to 80% of the time on the world's major telescopes. In many respects it is far more challenging to find appropriate imaging detectors for astronomical spectroscopy than for direct

249

imaging applications because the faint image of a distant object is made even fainter when dispersed into a spectrum. After turning things over in my mind it occurred to me that by allying conventional detector devices with an electronic processor and digital memory it would be possible to mimic the performance of a perfect detector. By analogy, I thought, the eye operates not only as an optical device incorporating an image sensing layer but also relies on the processing and memory functions of the retina and brain. In fact, by this augmented use of available photoelectronic imaging devices, it would be possible to detect individual photons of light and build up an image in a computer memory by accumulating photon counts in as many memory locations (therefore image elements) as needed for the application. At first I envisaged 10^3 image elements, then realized it was just as easy to achieve 10^6 or 10^7.

At very low light levels the ultimate limit to the accuracy of image recording set by the quasi-Poissonian statistics of the incident photons soon becomes evident. An ideal detector would unambiguously register the arrival of each photon in its correct image position, and a signal of N_1 photons per unit area would be measured with the maximum possible signal-to-noise ratio of $N_1/N_1^{1/2} = N_1^{1/2}$. Practical detectors invariably contribute additional noise fundamentally because a fraction of the photons goes completely undetected, but also due to uneven gain processes which accord detected photons unequal statistical weighting and to the addition of background noise. Additional losses of information in real detectors arise because of image spreading and halation effects resulting in loss of resolution and contrast. A frequently used figure of merit is the detective quantum efficiency (DQE), which may be defined as the ratio of the number of quanta N_1 required by an ideal detector to the number N required by the actual detector for identical output signal-to-noise ratio: i.e. $DQE = N_1/N$.

To overcome the deficiencies and limitations inherent in real image detectors the image photon counting technique shifts the emphasis from the detailed photometric performance of the devices themselves to the real-time data processing of the fundamental signals delivered by the primary photon-to-electron conversion layer in the system. The basic approach of the technique is to detect individual photon events in a two-dimensional image by means of a high-gain image intensifier optically coupled to a continuously scanning television camera (acting as a spatially sensitive one-frame buffer store) and to record their central positions in a digital electronic memory associated with an on-line computer.

Because an image photon counting system (IPCS) relies so heavily on the use of a high-gain image intensifier as its primary component I will spend a little time describing such devices. Then I will go more fully into the image photon counting technique itself.

2 Image intensifiers

All image intensifiers† used in astronomy depend on the initial conversion of photons into an electron image by the process of photo-emission. In one commonly used technique the electrons then are accelerated onto a cathodoluminescent phosphor screen to produce an enhanced optical image. A blue-light gain of 50 is typical for a simple device of this kind and several such stages can be arranged in cascade to give a very high overall gain. Means of focussing the electron image are shown schematically in Fig. 19.1. The simplest is proximity focussing, in which the electrons merely are accelerated between the plane-parallel photocathode and output screen which are narrowly separated (by about 1 mm) to limit the lateral spreading of electrons during transfer. The practical gain and resolution performance of proximity focussed intensifiers is limited by mechanical and electrical breakdown considerations and is lower than for types having electrostatic-lens or magnetic focussing. For highest gain and resolution the magnetic type is generally used. Cascaded stages are most simply produced by depositing the phosphor of one stage on one side of a thin transparent membrane (usually mica) and the photocathode of the next stage on the other side, and containing all stages in one vacuum envelope. A four-stage intensifier of this kind produced by Thorn-EMI has a blue-light gain of approximately 10^7. However, electrostatic lens focussing produces a more compact intensifier. In this case, for best electron-optical performance over the whole image area, both the photocathode and screen surfaces must be curved; this can present difficulties in matching optically, but the problem is usually overcome by using plano-concave fibre-optic faceplates to transfer light between outer plane optical surfaces and inner concave surfaces. Due to the curvature of the input and output fibre-optics several effects combine to give a radial decrease in light gain for such an intensifier, reaching about a factor 2 in multistage devices. In commercial devices a compromise is often struck in reducing the loss in centre-to-edge gain at the expense of off-axis

---

†The term intensifier here is used rather loosely: it implies a device with a net *photon* gain but with arbitrary input and output spectral characteristics; e.g. a device may accept an ultraviolet image but convert it to an enhanced optical image.

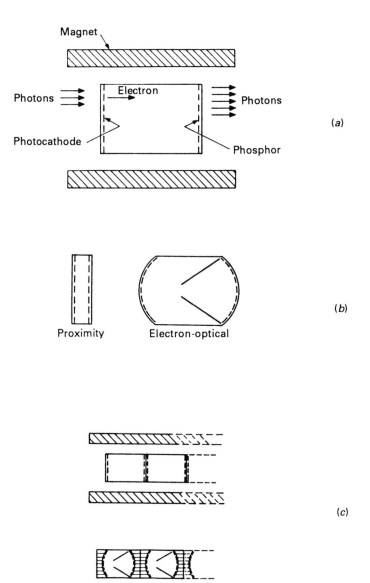

Fig. 19.1. Focussing methods and cascading of photocathode–phosphor image intensifiers: (*a*) magnetic focussing; (*b*) electrostatic focussing; (*c*) multistage assemblies.

resolution and the introduction of pin-cushion type distortion. On the other hand, magnetically and proximity focussed intensifiers have the advantage of plane electron-optical surfaces, not requiring fibre-optics, so have a more uniform response; also they can be made with an input window transparent further into the ultraviolet region than is possible for fibre-optic faceplates.

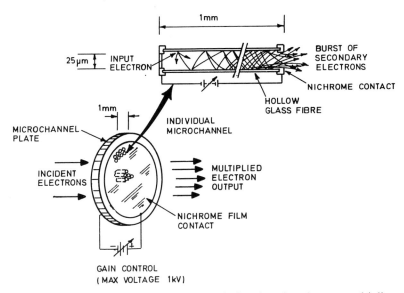

Fig. 19.2. The operation of a microchannel-plate imaging electron-multiplier.

The microchannel plate (MCP) is a different kind of imaging electron multiplier capable of extremely high gain within a small volume. It consists of a close-packed array of fine, slightly conducting glass tubes having a secondary emission coefficient greater than unity. An accelerated electron entering one of the channels and striking the wall produces secondaries which in turn are accelerated by an internal electric field again to strike the wall; the process continues until the avalanche of electrons emerges from the output of the channel (Fig.19.2). The MCP can be built into a very compact double proximity-focussed structure using a conventional photocathode on its input window to give a 'wafer' intensifier with high gain and no distortion (Fig. 19.3). Alternatively, with the omission of the input window and a suitable dielectric photocathode (e.g. CsI) deposited directly onto the channel walls, the device can be made sensitive from the far ultraviolet to the soft X-ray region. The electron gain of a typical MCP with straight channels is 10^4 at 1 k eV applied, normally limited by the onset of positive ion feedback, and the output pulse sizes exhibit a broad distribution which results in a noisy image and leads to a reduced DQE for the device. However, several techniques have been introduced to suppress ion feedback, for example to curve the channels or to stack two or three differently angled plates, and then an electron gain greater than 10^7 is achieved without difficulty. Additionally, the saturated output pulse sizes exhibited at such

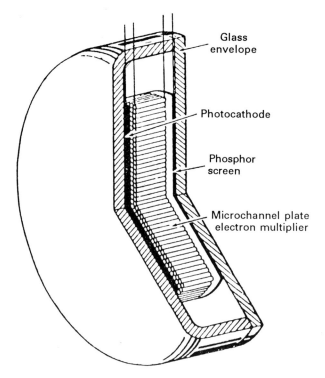

Fig. 19.3. Schematic representation of a 'wafer' intensifier (seen in cutaway section) incorporating a microchannel plate imaging electron multiplier.

high gains are confined to a sharply peaked distribution which substantially decreases the noise in the output image compared to the low gain performance. However, due to geometrical and other losses at the input to the multiplier a substantial fraction (perhaps $\frac{2}{3}$) of the signal electrons may be lost before a detectable avalanche is produced. The loss generally is greater than the corresponding loss in photocathode–phosphor intensifiers.

3 Image photon counting systems

The basic method of the technique of image photon counting is to amplify the photoelectron image by means of a suitable image intensifier to a level where all the scintillations from individual detected photons can be discerned above the readout noise of a television camera used to view the output screen of the intensifier. In the original IPCS which I developed with colleagues at University College London, a lead oxide vidicon (Plumbicon) is lens-coupled to a four-stage magnetically focussed intensifier and a signal of several hundred thousand electrons ultimately is generated in the camera tube target for each detected photon. The video output contains, in addition to the photoelectron scintillation information, high frequency noise and low frequency level changes. A pulse amplifier with optimized integration and differentiation time constants is used to reject much of the noise. The pulses are fed to a hard-wired digital signal processor, and thence to a computer. The special processor is essential for a true photon counting system; its principal functions are:

(a) Location of the geometric centre of each event ('centroiding'). This is necessary because the finite size of a scintillation means that it is detected on perhaps two or three adjacent television scan lines. Centroiding thus ensures both maximum DQE by according equal statistical weight to each counted event and markedly improved spatial resolution. The latter is illustrated in Fig. 19.4, the upper half of which represents the progression of the signal from an individual detected photon through the IPCS. A photoelectron released from the input photocathode (A) may deviate slightly from the axis according to the magnitude of its emission velocity components which are statistically distributed. The primary photoelectron therefore lands on the first electron-multiplying element (B) (e.g. a phosphor–photocathode dynode) at a position determined by a two-dimensional probability function; the size and shape of this function are identical to the large-signal 'point-spread function' (PSF)[†] which is generated by the statistical accumulation of individual electrons (ΣB). The slightly divergent beam of secondary electrons from the point of impact of the primary on (B) passes stage by stage through the rest of the system and gives rise to a scintillation of finite diameter on the camera tube target (C). The centroiding logic determines the geometric centre of the scintillation and this is recorded at the appropriate address in memory (D) ideally corresponding to the point at (B) on which the primary photoelectron was incident. For a large number of electrons originating from a point image at (A), integration in the external memory (ΣD) results in a PSF similar in size and shape to the probability function at (B). By

---

[†] The PSF is the shape of the output image resulting from a point input; its diameter at half maximum height may be regarded as a measure of the loss of resolution in the imaging system.

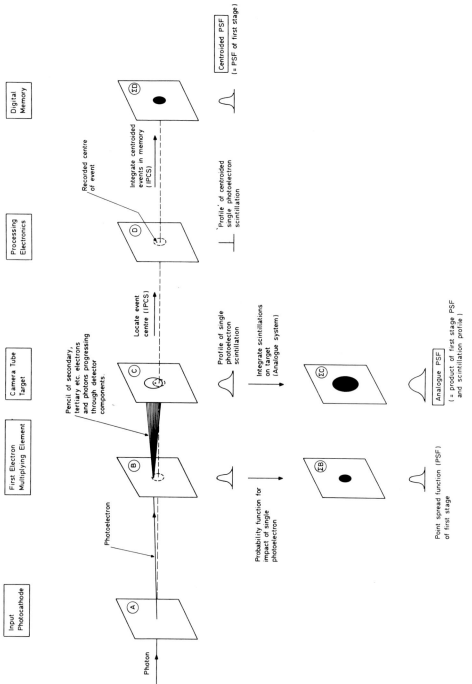

Fig. 19.4. Illustration of the progression of the signal from an individually detected photon through the IPCS.

contrast, in an analogue system the signal from each photoevent is not processed independently; the PSF of the first stage (Σ B) is further broadened (by convolution with the scintillation profile) to give a larger output PSF at the camera tube target (Σ C). Thus, there is considerable resolution loss in an analogue system compared to a centroiding IPCS.

(b) Rejection of low-amplitude noise pulses from the signal amplifier, and of high-amplitude pulses due to spurious ion scintillations in the intensifier. For efficient discrimination, it is only necessary to have adequate gain and for the scintillations to have a favourably peaked brightness distribution, as is obtained with photocathode–phosphor intensifiers or with MCP intensifiers operating in the pulse-saturated mode.

(c) Encoding the positions of the centroided photon events and passing the data to the memory where the appropriate addresses are incremented.

Because of the digital (photoevent/no photoevent) mode of operation, the system performance is far less sensitive than in the analogue case to detailed aspects of the detector's construction; the performance on an IPCS is generally limited only by the efficiency of the primary photocathode and by the resolution of the first intensifier stage.

The separation of the detection and storage functions of an IPCS results in several additional advantages:

(i) Storage capacity is effectively unlimited, being completely independent of the detector and depending only upon the size of the memory.

(ii) Rapid time-varying effects can be accommodated.

(iii) There is no low level threshold for very weak exposures so that very faint images with only a few photoelectrons per pixel can be recorded with maximum DQE.

(iv) Real time display of the accumulating image on a monitor results in high operating efficiency; considerable observing time may be saved by terminating exposures as soon as the required image signal-to-noise ratio has been achieved.

(v) The exposure linearity and system stability allow accurate photometric calibration and background subtraction. The latter is particularly useful in astronomical spectroscopy of faint objects, where the signal can be extracted from a background

Fig. 19.5. The IPCS detector head unit, containing image intensifier, coupling lens and television camera.

containing a bright night-sky continuum and emission lines; subtraction and calibration may also be carried out in real time.

(vi) Within a limit of about 10^7 available pixels the data acceptance format is completely flexible; the system may be pre-programmed to accept image information from selected areas, or to sum over adjacent groups of pixels.

A picture of the IPCS detector head unit is shown in Fig. 19.5, and an electronic block diagram of the initial common-user system, with representation of the signal processing functions, in Fig. 19.6. Not shown here is a facility, added later, for internally scanning the whole image format laterally within the intensifier to smooth out granularity introduced by the first, second and third phosphor screens. A display of raw instantaneous photon events in a faint image as detected in a part of one TV frame is shown in Fig. 19.7. An indication of the gain in spatial resolution achieved when using the electronic centroiding technique as shown by the widths of comparison arc lines obtained during spectrographic recording is given in Fig. 19.8.

The IPCS has been in routine observing use at several observatories since 1973. A system similar in configuration to the UCL IPCS is to be used on the Hubble Space Telescope. Examples of astronomical results obtained with the system are given in Fig. 19.9.

Several other image photon counting (or quasi-counting) detector systems have been developed using various forms of readout including solid state, resistive and other position-sensitive layers. Some have more restricted channel capacity than the IPCS but are excellent for many one-dimensional spectroscopic applications. One of these, the Digicon, developed by Beaver and McIlwain, which found initial use for ground-based optical astronomy, also is to be used on the Hubble Space Telescope.

Further developments involve the use of a CCD for readout of the photon

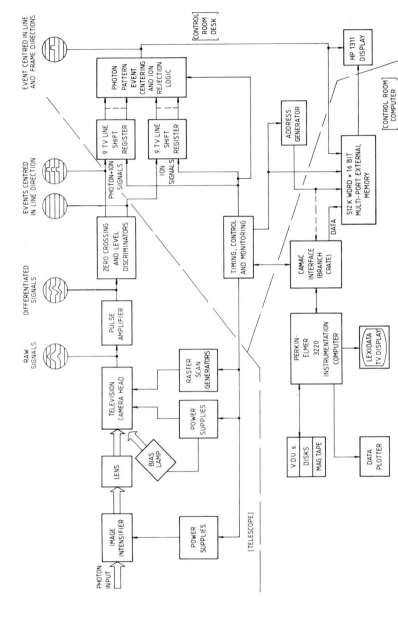

Fig. 19.6. Block diagram of the IPCS as installed on the Isaac Newton Telescope at La Palma, showing the electronic processing functions.

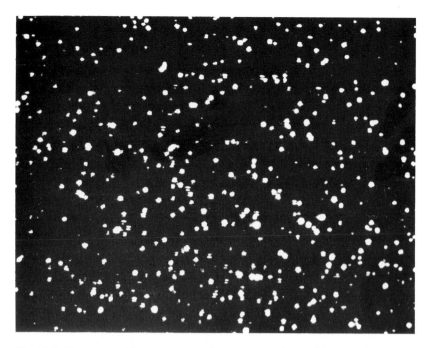

Fig. 19.7. Unprocessed instantaneous photon events detected in a small part of
the TV frame by the IPCS detector head unit. The events in general span
more than one TV raster line.

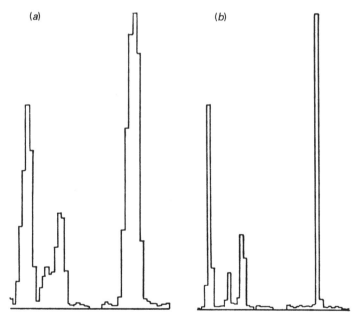

Fig. 19.8. Small region of a comparison arc spectrum recorded with the IPCS.
The centroided line-spread function (b) is clearly less than one channel (25 μm
in this case), while the non-centroided case (a) extends over many channels.

information. One such system has been developed by Stapinski and others at Mount Stromlo and Siding Spring Observatories, using a CCD optically coupled to a MCP intensifier. A similar system, using a reducing fibre-optic feed from the MCP to the CCD and also lens coupling to a standard EMI intensifier, has been developed jointly by University College London and the Royal Greenwich Observatory; an additional feature of this system is electronic interpolative centroiding which effectively increases the number of available image elements by a factor 64 over what the CCD itself can provide.

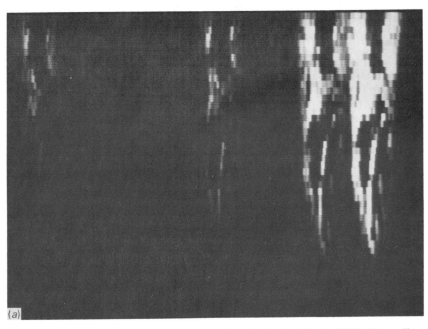

Fig. 19.9. Examples of astronomical results obtained with the IPCS: (*a*) small region of a two-dimensional slit spectrum (the vertical dimension is along the slit) of the Crab Nebula prominently showing (right to left) the [OIII] 5007, 4959 Å and Hβ emission lines indicating expanding motion (obtained by P.G. Murdin on the Anglo–Australian Telescope); (*b*) one-dimensional spectrum of a QSO showing red-shifted intrinsic broad emission lines (the prominent one is the Lyman α line of hydrogen at 1216 Å in the rest frame) and sharp absorption lines due to intervening material, with the total 1σ photon noise in the signal indicated at the bottom (obtained by A. Boksenberg and W.L. Sargent on the Palomar 5-m telescope); (*c*) data 'cube', acquired by the scanning Fabry Perot inteferometer system known as TAURUS, of the galaxy NGC 1365 in the region of the [OIII] 5007 Å emission line, showing the conventional two dimensions at middle, one spatial dimension and line velocity (vertical) at bottom and line profiles at spatially selected positions at top. (Obtained by K. Taylor on the Anglo-Australian Telescope.)

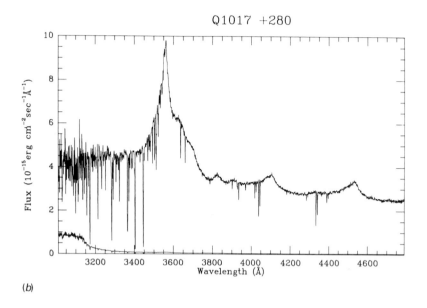

(b)

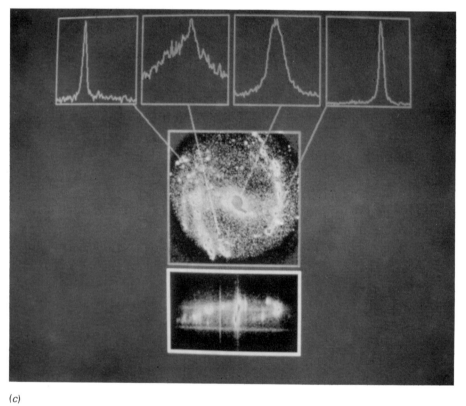

(c)

Fig. 19.9. (*Contd.*)

4 Conclusion

Improvements in image detection are likely to stem at least as much from the application of new electronic processing and data reduction methods as from the development of better devices. While there remain many applications in astronomy for image photon counting systems the CCD used as the primary detecting element is taking an increasing role, and in future all detector devices may be of this type. Means of sensitizing them to operate efficiently in the ultraviolet already exist, and internal gain might be achieved by avalanche processes already implemented in discrete diode structures. The technique of image photon counting, with its superior low light level performance, may then be achieved entirely in the solid state.

20

Developments in solid state detectors

C.D. MACKAY

Summary

A brief review of the possibilities for solid state detectors in optical/infrared astronomy is presented.

Avalanche photodiodes

Before discussing solid state arrays I want to draw attention to the use of avalanche photodiodes as single element light detectors. They have several attractive features as potential replacements for photomultiplier tubes in many applications.

When an avalanche photodiode is biassed above breakdown a large current flows. If this is limited to a few tens of microamps the current is unstable and can switch itself off. It will stay off until another avalanche breakdown is triggered by a carrier arising in the avalanche region either from internally generated dark current or because of a photo-generated carrier. Even with fairly simple circuits photon counting rates of several megahertz can be obtained. Devices are manufactured by Hammamatsu (Japan), NEC (Japan), RCA (USA) and others. They are available in chip form (see Fig. 20.1), TO-18 encapsulation and with integrated optical fibre pigtails (Fig. 20.1). Their properties as photon-counting devices are excellent (Brown, Ridley & Naritz, 1986). They have high peak responsive quantum efficiencies (RQE), 70% at 850 nm (silicon, RCA), 1.5 μm (germanium, NEC) and 1.6 μm (indium gallium arsenide, NEC). In practice the maximum detective quantum efficiency (DQE) that can be achieved is only about half the RQE value. Their principal disadvantage is that they need to be cooled to

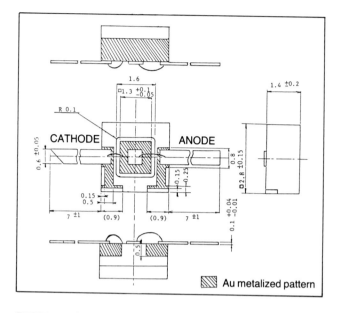

Au metalized pattern

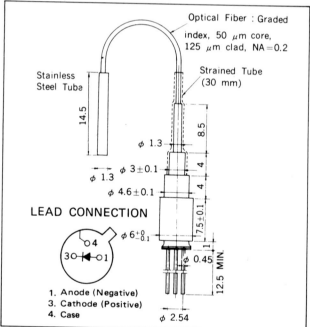

Optical Fiber : Graded
index, 50 μm core,
125 μm clad, NA=0.2

Strained Tube
(30 mm)

Stainless
Steel Tube

LEAD CONNECTION

1. Anode (Negative)
3. Cathode (Positive)
4. Case

Fig. 20.1. Avalanche photo diode: typical example.

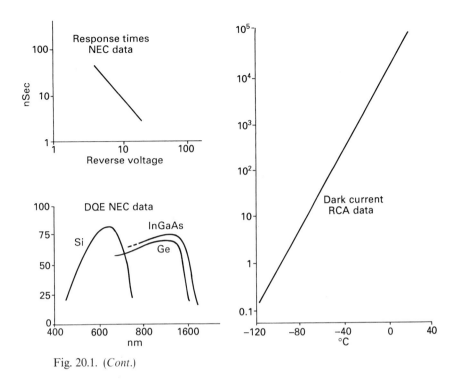

Fig. 20.1. (*Cont.*)

the sort of temperature used for CCD systems ($-80\,°C$ to $-100\,°C$) to suppress their dark current (Fig. 20.1). However their robustness and insensitivity to damage from light overload ensures they will be increasingly used by astronomers. Typical costs are $70 for small silicon devices, $300 for germanium and $2500 for InGaAs, presumably decreasing with time.

Charge coupled devices

The most widely used solid state detectors are charge coupled devices (CCDs). Their general characteristics have been reviewed recently (Mackay, 1986). Unfortunately the future outlook is much less certain than it seemed only a couple of years ago. Two major US CCD companies (RCA and Texas Instruments) have recently withdrawn from CCD manufacture. In Japan the manufacturers prefer device architectures they believe best for colour imaging devices for domestic use. These are quite unsuited to astronomers needs. In Europe things look much better. EEV/GEC Ltd (UK) have recently announced a new range of devices and cameras and their frame transfer devices are fairly widely used in astronomy. Thompson CSF (France) and Philips (Netherlands) both manufacture a range of frame transfer CCDs that are probably suitable for astronomical application. More recently in the US developments have come from Tektronix and Reticon. Tektronix have experienced manufacturing problems and use a device structure ideally suited to astronomers' needs but not suitable for more conventional applications at TV frame rate, an aspect that must reduce their potential market and therefore its longer term availability.

As far as future performance is concerned we can expect improvements in cosmetic quality, blue response, overall DQE (see Mackay, 1986), and read-out noise–probably approaching 1 electron rms. However it is likely for many ultra-low light level applications that the main limit to their use will be charge-transfer efficiency at low light levels. Unfortunately this is something only slightly understood by CCD designers and not of great concern in most other applications at higher light levels. For these reasons manufacturers cannot be expected to put so much effort into improving it to the levels astronomers would like to see.

What of the prospects of larger devices? The development of 1500×1500 pixel CCDs by GEC produced several working devices from the small number of batches produced on a best-efforts basis for the Anglo–Australian Observatory. The transfer of CCD production from GEC to EEV has halted this programme pending the outcome of the Tektronix effort to make thinned 2048×2048 devices. Even if Tektronix can make them, the high price tag ($75 000$) is going to limit their use to a small number of very well heeled observatories. Indeed, it is not even clear that such large devices are really what astronomers need. It is inevitable that they will be cosmetically poorer than small devices and have charge transfer problems at least four times worse since the charge has to be transferred four times as far. It is also going to take many minutes to read out the 4 million pixels. There has always been a significant number of astronomers who always clamour for the biggest (detectors, telescopes, computers). However, it is interesting to note that although arrays of smaller CCDs have been practicable for several years with much less effort or expense than the big devices entail, nobody has attempted to use two or more simultaneously for any real application.

In the future astronomers may find that they have to invest more in the manufacture of CCD arrays to their own specification. There are a number of trends to make CCDs better suited to domestic TV frame rate applications that will compromise their performance for astronomy. Astronomers may have been very fortunate in that the standard devices made by several companies have been well matched to their requirements. In the future they may well have to invest more to push developments their way.

Infrared arrays

The situation with IR arrays is much less satisfactory for a number of reasons. Visible CCDs were developed as imagers for a competitive commercial market and so are rather good value for money. IR arrays are being developed for military applications where the device costs are not subject to any commercial control and so are generally rather poor value for money. Further, the interest of the military in the devices means that astronomers are lucky to get access to technologies that are, at best, medieval and not at all representative of state of the art. It is remarkable that some of the best IR images have recently come from the Rochester group using a 32×32 array manufactured in 1974.

The emphasis on military development programmes is on high device uniformity and minimum cooling rather than the high DQE and low read-out noise needed by astronomers. It is also the case that many of the best IR detectors are often integrated with CCD read-out multiplexers that are of remarkably poor quality.

Astronomers have become involved in state-of-the art technologies in many areas of importance to the subject. Mirror technology, optics, computers and image processing software are areas where the technical involvement of the astronomical community has been valuable and productive. It may be that it will soon be necessary for the involvement to extend to solid state detector development, particularly in the IR where there is most to be gained. Such work would have to be carried on outside the US where export restrictions presently forbid the sale of devices made with 10-year-old technologies of negligible strategic importance. It has to be true that the returns that would come from investing a few million dollars into such programmes would dwarf the returns from the much larger sums presently being devoted to the next generation of ground based telescopes. If we simply wait for something to turn up from the downgraded classification of some obsolescent military detector development pro-gramme we cannot expect our requests for the next generation of telescopes to be taken as seriously as they should be.

References

Brown, R.G.W., Ridley, K.D. and Naritz, J.G. (1986). *Appl. Optics.* In press.
Mackay, C.D. (1986). *Ann. Rev. Astr. Astrophys.*, **24**, 255–83.

21

The use of CCDs on optical telescopes

P. R. JORDEN

Summary

An introduction to the main characteristics of CCDs (as used for optical astronomy) is given, together with a review of the types of device available for use. In subsequent sections, certain important characteristics of this detector are described and illustrated with examples; spectral response, dynamic range and detector formats are discussed. The variety of optical astronomy applications is indicated, with references to examples of many types of use. Finally, current performance limitations are discussed, and this leads to an examination of expected developments in the near future.

1 Introduction

Charge-coupled devices (CCDs) have been widely adopted for use on optical telescopes over the last five years, and were first used nearly ten years ago. These silicon array detectors have enabled astronomers to see deeper, and at redder wavelengths than was previously possible. Here, I shall discuss the particular advantages of this type of detector for optical astronomy, with illustrations from a variety of typical applications.

The operational parameters and detection characteristics will be introduced in order to indicate the advantages, and limitations, of the CCD as an optical detector. As is usual with any detector we do not have the perfect device, but the limited number of arrays from a few manufacturers do show us that some of our observational demands can be achieved. Some indications of future trends and improvements will be given.

At the observatory on La Palma we have a variety of CCD cameras in

271

regular operation; examples of data from these instruments will be used to demonstrate the use of CCDs for optical astronomy. It is clearly impossible to give a complete review of the applications of these detectors, but some of the most common uses will be illustrated (see also Jorden, Thorne & Waltham, 1986).

2 Detector characteristics

The main advantages, and some limitations, of the CCD as a low light level detector derive from its construction as a two-dimensional silicon array. We achieve high efficiency, direct conversion of photons to electrons, and can integrate the signal for long periods. However, the format of available detectors is determined strongly by the prior development of

Table 21.1. *CCDs for astronomy*

Manufacturer	Format	Availability	Comments
Fairchild	100 × 100 380 × 488 + others	< 1980–present day	Low QE high noise not widely used now
RCA	320 × 512	< 1980–1985 no longer made	Thinned, high QE. High read-out noise
GEC/EEV	385 × 578	< 1980–present day	Low readout noise, widely used
Thomson-CSF	384 × 576	~ 1983–present day	Low noise, cosmetically clean
Philips	604 × 576	~ 1984–present day	Unusual structure, not used yet
Texas Instruments	800 × 800 + others	< 1980–present day	Not widely available outside of USA
Textronix	512 × 512 2048 × 2048	mid 1986–1987?	Only just available. Promised for 1987
Hitachi	590 × 582	~ 1985	Cameras only available
Sony	?	?	Little information available
Kodak	> 1000 × 1000	?	May be of no use for astronomy
Fuji-film	> 1000 × 1000	?	"

Table 21.2. *CCD characteristics*

Good quantum efficiency	– up to 80%
Good red-wavelength response	– λ up to 1.1 μm
Finite readout noise	eg $\sigma_r \sim 5$ electron
Good dynamic range	10^4 typically
Compact and stable (spatially and temporally)	
Typical format	385 × 576 pixels (15–30 μm)

TV-format sensors (eg 380 × 580 pixels) and by restrictions of the semi-conductor industry fabrication capabilities. A summary of commercially available CCDs is given in Table 21.1

The primary advantage that the CCD has over other two-dimensional imaging detectors is that it achieves high quantum efficiency, and its response extends to beyond 1 μm. An important disadvantage of the CCD is that it has a finite readout noise, typically of 5–10 electrons rms; this is the random fluctuation of signal that is seen on every pixel-readout, even at zero signal level. This precludes its use for direct photon-counting and determines the s/n at low light level.

A large dynamic range (of order 10^4) is achieved by this type of detector; this is the ratio of the pixel full-well capacity to the readout noise. For most typical CCDs a full-well of about 10^5 electrons is obtained; this is the maximum charge at which a linear response is still seen. This large, linear, response range, particularly from a stable detector, enables us to record certain astronomical objects better than before.

The compact nature of this solid-state detector has allowed some rather novel and powerful instruments to be built. A good example is the Faint Object Spectrograph (FOS) that utilizes a CCD within a form of Schmidt camera for very efficient low-dispersion spectroscopy.

The main CCD characteristics are given in Table 21.2; some of these will then be illustrated in the following sections. A further discussion of specific CCD characteristics can be found in Thorne *et al.* 1986.

2.1 Quantum efficiency and wide spectral response range

The silicon CCD exhibits a large spectral response range, the long wavelength cut off being nearly 1.1 μm. The short wavelength response depends more strongly on the design of the device but can extend to

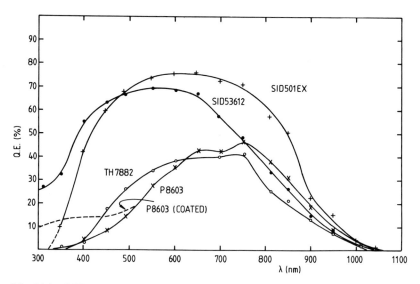

Fig. 21.1. CCD quantum efficiency. Curves for normal (thick) devices, (GEC and Thomson) and for thinned devices (RCA) are shown. The response of a dye-coated CCD is also indicated (see text later).

considerably lower than 300 nm. We can therefore cover the normal photographic and photocathode wavelength ranges as well as the previously unexplored near-infrared region. Fig. 21.1 indicates CCD quantum efficiency.

As an illustration of the far-red response, Fig. 21.2 shows a spectrum of the planetary nebula NGC 7027 centred at 1.05 μm. This was recorded on the 2.5 m Isaac Newton Telescope (INT) Cassegrain spectrograph in 1000 sec with the GEC P8603 CCD. A variety of astrophysically interesting emission lines is to be found near the 1 μm wavelength. The 1083 nm He I line is clearly seen from this bright nebula and we can also detect the 1094 nm HI line. The FOS also utilizes the whole wavelength range of the CCD and this is described in a later section.

The CCD is also widely used for multi-band direct imaging, particularly for photometry. A good example of this application can be seen in Fig. 21.3. The X-ray source field GX 354 + 0 was recorded at two red wavebands R and Gunn-Z, (centred on about 700 and 950 nm respectively). The ability to see faint, heavily reddened objects is illustrated by the detection of many more stars with the Gunn-Z filter, despite the reduced CCD response at this wavelength.

A further demonstration of the wide spectral sensitivity range is to be seen

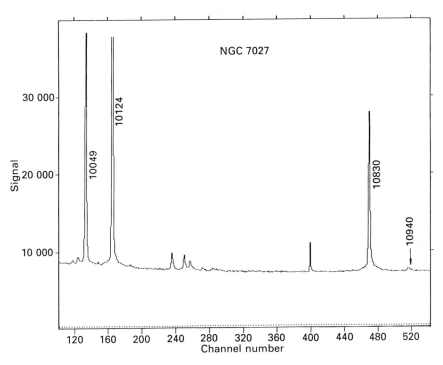

Fig. 21.2. Far red spectrum of NGC 7027, covering 1000–1100 nm.

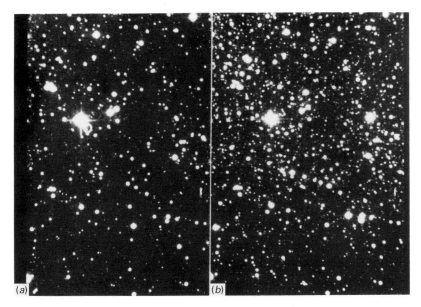

Fig. 21.3. The X-ray source GX354 + 0, (*a*) at R band, (*b*) Gunn-Z filter.

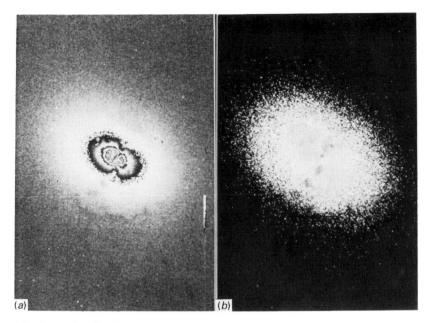

Fig. 21.4. The elliptical galaxy Fornax A, (*a*) broad band B image, (*b*) colour-subtracted B-GunnZ image.

in Fig. 21.4. The elliptical galaxy Fornax A (NGC 1316) was observed with an RCA CCD at a variety of broad-band wavelengths, including B (450 nm) and Gunn-Z (950 nm). The direct blue image is seen in 21.4(*a*) and a colour-divided frame (B-GZ) is shown in 21.4(*b*); the enhanced detail of the reddened dust-lane is clearly shown in the latter image. This type of image processing is particularly effective on stable devices with wide dynamic range such as the CCD (see also Carter *et al.*, 1983 and Sparks *et al.*, 1985).

2.2 Dynamic range

The intrinsic dynamic range of most CCDs exceeds 10^4 and can approach 10^5; the linear response of this integrating detector also allows an extension to the available photometric range since exposure times from < 1 sec to 3000 sec or so are practical.

A particular type of observation for which the CCD is well suited is the measurement of an intense (compact) object surrounded by a much fainter (extended) one. Fig. 21.5 shows such an example; the galaxy PKS 0244-304 can be seen to have an inner spiral of high intensity; there is also a much

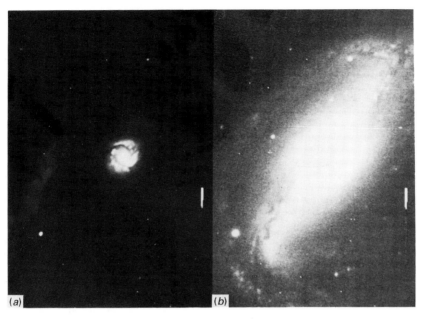

Fig. 21.5. A single CCD exposure of PKS 0244-304, reproduced as two frames. (a) The bright core detail is shown, (b) The fainter extended structure is displayed (with the core shown as a photographic emulsion would record it).

larger, fainter spiral structure. The measurement of a bright core simultaneously with a diffuse region is not practical by photographic techniques. The discrimination of a bright quasar from an underlying galaxy is a similar example (see Gehren *et al.*, 1984).

The measurement of small absorption features on a strong continuum background poses a problem for an instrument such as the IPCS with a more limited dynamic range. Thus the CCD, although not able to detect individual photons is able to do this sort of spectroscopy rather well. In this case, as well as direct imaging to faint limits, it is very important to obtain a good flat-field calibration.

2.3 Compact detector format

The compact form of a CCD, with its direct electronic readout allows certain novel instruments to be constructed; a good example is the Faint Object Spectrograph (FOS) shown in Fig. 21.6. This instrument uses the CCD in an enclosed, and cryogenically cooled, Schmidt camera

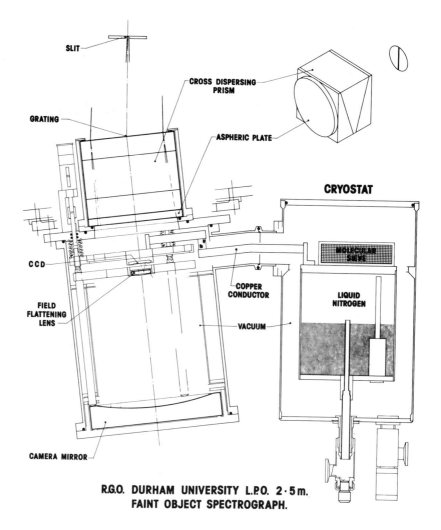

**R.G.O. DURHAM UNIVERSITY L.P.O. 2·5m.
FAINT OBJECT SPECTROGRAPH.**

SLIT

CROSS DISPERSING PRISM

GRATING

ASPHERIC PLATE

CRYOSTAT

CCD

COPPER CONDUCTOR

LIQUID NITROGEN

FIELD FLATTENING LENS

VACUUM

CAMERA MIRROR

SPECIAL FEATURES OF FAINT OBJECT SPECTROGRAPH.

1. High efficiency :-

   > No collimator
   > Few optical surfaces
   > Minimum vignetting.

2. High quantum efficiency CCD detector.
3. Wide spectral range (350 - 1100 nm.)
4. Fixed format for easy on-line data reduction.
5. Cross dispersed 2 order or long slit single order operating modes.
6. Uses existing spectrograph slit assembly.
7. Allows quick interchange or even simultaneous operation with intermediate dispersion cameras.

Fig. 21.6. The Faint Object Spectrograph.

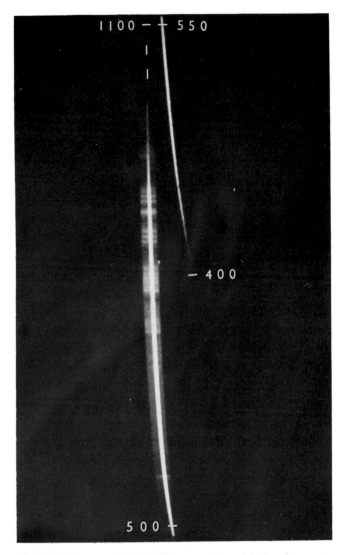

Fig. 21.7. An example of FOS image; 1st and 2nd order spectra can be seen.

configuration with a prism and transmission grating (grism). The result is a very efficient, fixed-format low dispersion spectrograph which records a spectrum utilizing the whole wavelength range of the CCD. (The device is described fully by Breare, Ellis & Powell, 1986.) Fig. 21.7 indicates the form of the 1st and 2nd order spectra obtained with this device.

The throughput of the spectrograph is very high and therefore allows

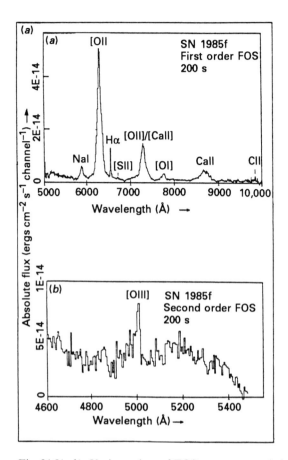

Fig. 21.8(a, b). Various plots of FOS spectra recorded on the INT; the wavelength response of the instrument, as well as the strong night-sky lines can be seen.

redshift surveys of faint quasars and similar measurements. It is found that the accuracy of sky-subtraction is an important factor in determining the final s/n; in principle, the stable detector allows this to be done well, despite the many strong sky emission lines that are seen. Fig. 21.8 gives some examples of recorded FOS spectra obtained on the INT.

It is also intended to use the CCD as the basis for the telescope autoguider systems currently being designed for the 4.2 m William Herschel Telescope. Again the compact imaging detector is advantageous, although a Peltier-cooler is necessary to reduce the dark current. The robustness and stability of the detector will be very useful in this application.

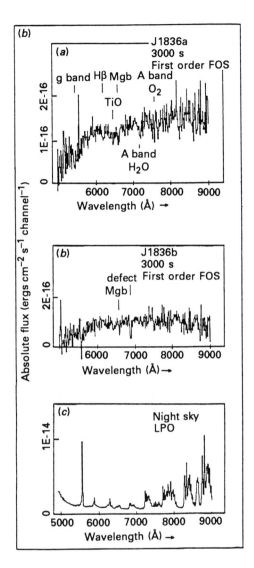

Fig. 21.8. (*Cont.*)

3 Optical astronomy applications

The previous examples have shown some of the uses of CCDs in astronomy; Table 21.3 summarizes the main application areas, with references to an example of each.

We have not discussed photon counting here, but the CCD is available as an alternative to other TV cameras or detectors, when preceeded by an

Table 21.3. *CCD applications*

		References
Direct imaging	Photometry (of compact sources)	(Walker, 1985)
	Surface photometry	(Sparks *et al.*, 1986)
	Faint detections	
Spectroscopy	Low resolution (FOS)	(Lawrence *et al.*, 1986)
	Intermediate dispersion	(Diaz, Bagel & Terlevich, 1985)
	High dispersion (e.g. Echelle)	(McKeith *et al.*, 1987)
Photon counting	used with an intensifier	(Fordham, Bone & Jorden, 1986)
Scanning mode		(McGraw, Cawson & Keane, 1986) (A CCD/Transit Instrument)
Autoguiders or	star-trackers	(Lange, Mossbacher & Purll, 1986) (The ROSAT star tracker)

image intensifier. The main advantage is its stability; however, the small format and limited readout rate can be a restriction.

The CCD can be used in a scanning-mode whereby the projected image is mechanically scanned in synchronism with the electronic readout; this can be used to minimize pixel–pixel sensitivity variations, but has also been used to implement a rather elegant transit sky-survey programme.

Although it does not fall within the scope of this paper, it should be pointed out that CCDs are used extensively for a variety of other science applications. X-ray astronomers use the ability of the CCD to give energy and spatial resolution of detected electrons generated by X-rays (Lumb, Hopkinson & Wells, 1985). Particle physicists similarly use the CCD, usually in large detector arrays, to analyze high energy physics interactions.

4 Performance limitations

So far, various positive features of CCDs have been presented and in this section some of the disadvantages or current limitations of performance will be discussed.

Readout noise needs to be low, particularly for spectroscopic (low background) applications. Much astronomy can be done with a noise level of order 10 electrons; however, we must still use image intensifiers (with their lower quantum efficiency) in order to do photon-counting. It is expected that manufacturers will continue to improve their devices to achieve noise levels of order 3–5 electrons more commonly. There is also a good possibility that different on-chip amplifiers can be constructed to approach the 1 electron readout noise.

Quantum efficiency is already very respectable for the silicon CCD; however, most of the currently available devices have a peak QE of 40%. It is possible to produce a thin, backside-illuminated device with a peak QE of about 80%; this also has a much extended blue-response. We must hope, and encourage, the principal manufacturers to develop their CCD-production in order to give us this extended response. (RCA achieved good results, but did not make a low-noise device, and no longer make CCDs.)

A dynamic range of 10^4 has been quoted earlier and this is sufficient for a wide range of astronomical projects; the associated linearity characteristics are equally important. Most CCDs can be fabricated, and operated, so as to give a very linear response over the whole of their dynamic range. However, some CCDs exhibit a so-called 'fat-zero'; this is a threshold of sensitivity which prevents low-level signals from being measured unless a pre-flash exposure of uniform illumination is given.

Although a preflash pattern of, say 200 electrons, can be subtracted to high accuracy the residual contribution of photon shot noise means that the effective readout noise is increased. Thus, a very low noise CCD is only useful if no sensitivity threshold is present. This threshold is due to internal charge-trapping and is a combination of CCD design and fabrication parameters. CCDs can be made with no threshold, but many suffer this at present.

Cosmetic quality is a term which astronomers use in a different way to the TV or CCD manufacturers. For normal TV uses it is important that the subjective appearance is very good; this results in a 'blemish' being defined as a point where the response is 15% below normal. The aim is to have no blemishes. For astronomy, we operate over a much wider signal range and could tolerate one or two fixed blemishes provided that good response was seen, at low light level, elsewhere. There is thus a difficulty of specifying and testing devices; this is being met by suitable communication with the manufacturers so that they understand our problems! The remaining aim of achieving 'perfect' cosmetic quality is slowly being approached as silicon manufacturing technology improves.

The most common format of CCD is about 385 × 576; devices of this size can be fabricated with reasonable yield and are useful as TV-format sensors. The pixel sizes of 15–30 μm are usually appropriate for current applications, although there are clear cases where different sizes are desirable. The principal restriction is that of total size; we need CCDs of length 40 mm or so to utilize the full spectral range of typical spectrographs or to give better coverage of the sky for direct imaging. A few manufacturers have attempted to make large-format detectors (> 1000 × 1000); none are commercially available yet.

5 Looking into the near-future

The disadvantages imposed by the existing small size of detectors can be overcome through the use of large-format devices. Texas Instruments have demonstrated 800 × 800 devices (mainly used in the USA) and GEC have made some experimental 1500 × 1500 devices (under contract to the AAO). Tektronix have announced a 2048 × 2048 device which it is hoped will be available soon; some prototypes have been made but production difficulties have prevented manufacture during 1986. It is clear that the manufacture of such large-area silicon detectors is technologically difficult and the yields are low (often zero!). As an illustration of the use of such a large CCD Fig. 21.9 shows an image of the moon taken on the RGO 36in telescope; the area covered by a small-format GEC detector is shown for comparison.

Another way of achieving a larger detector area is by mounting an array of small CCDs at the focal plane. Use of standard devices invariably means that the image-area dead-space is quite large due to the finite package size. However there are indications that several manufacturers (including GEC and Thomson-CSF) could mount several CCDs in a line prior to final packaging. Such a device gives an extended area which is intermediate between true large-format devices and normal-format ones; the production yield and therefore cost would be realistic. It also appears possible to achieve a very small dead-space (< 1 pixel), although this has not yet been demonstrated at low cost.

Quantum efficiency has been discussed earlier; at present there are very few high-efficiency CCDs available. The process of thinning the CCD to about 15 μm, to achieve high efficiency through backside-illumination, is very difficult. Tektronix have advertised this type of device but are suffering

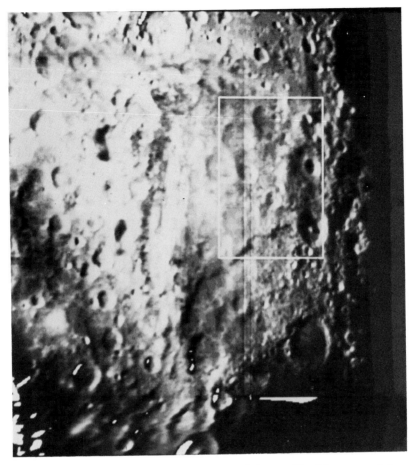

Fig. 21.9. An illustration of a large-format CCD image. The Moon recorded on the RGO Yapp telescope with a GEC MA703 device.

production difficulties; both GEC and Thomson-CSF aim to produce such devices in the near future. The technique is certainly feasible; the difficulty is to develop the production methods to achieve good results. An alternative method for achieving finite sensitivity at blue and near-UV wavelengths consists of a thin surface coating of UV-dyes on the CCD. This technique has been demonstrated to show good promise (Cullum *et al.*, 1986) although there are some doubts outstanding about its long-term stability.

In conclusion, it is clear that CCDs will be a dominant type of detector for many years. The role of photon-counting detectors in astronomy will remain, at least for the near future. CCDs will not totally replace other devices, including photographic plates, which will continue to have their scientific applications.

Acknowledgements

Thanks are due to Dave Thorne for contributing some material, and discussions; and also to Bill Martin for providing some references.

References

Breare, J.M., Ellis, R.S. & Powell, J.R. (1986). *Proc. SPIE*, **627**, 273.

Carter, D., Jorden, P.R., Thorne, D.J., Wall, J.V. & Straede, J.O. (1983). *Mon. Not. R. Astr. Soc.*, **205**, 377.

Cullum, M., Davies, S., D'Odorico, S. & Riess, R. (1985). *Astr. Astrophys.*, **153**, L1.

Diaz, A.I., Pagel, B.E.J. & Terlevich, E. (1985). *Mon. Not. R. Astr. Soc.*, **214**, 41P.

Fordham, J.L., Bone, D.A. & Jorden, A.R. (1986). *Proc. SPIE*, **627**, 206.

Gehren, T., Fried, J., Wehinger, P.A. & Wycoff, S. (1984). *Astrophys. J.*, **278**, 11.

Jorden, P.R., Thorne, D.J. & Waltham, N.R. (1986). *Proc. ESO/OHP Workshop on CCDs*, June 1986.

Lange, G., Mossbacher, B. & Purll, D. (1986). *Proc. SPIE*, **627**, 243.

Lawrence, A., Walker, D., Rowan-Robinson, M., Leech, K. & Penston, M.V. (1986). *Mon. Not. R. Astr. Soc.*, **219**, 687.

Lumb, D., Hopkinson, G. & Wells, A. (1985). *Adv. E.E.P.*, **64**, 497.

McGraw, J.T., Cawson, M.G.M. & Keane, M.J. (1986). *Proc. SPIE*, **627**, 60.

McKeith, C., Bates, B., Catney, M., Barnett, E., Jorden, P.R. & van Breda, I.G. (1987). *Astr. Astrophys.*, **173**, 204.

Sparks, W.B., Wall, J.V., Thorne, D.J., Jorden, P.R., van Breda, I.G., Rudd, P.J. & Jorgensen, H.E. (1986). *Mon. Not. R. Astr. Soc.*, **217**, 87.

Thorne, D.J., Jorden, P.R., Waltham, N.R. & van Breda, I.G. (1986). *Proc. SPIE*, **627**, 530.

Walker, A.R. (1985). *Mon. Not. R. Astr. Soc.*, **214**, 45.

22

On-line data handling

P.T. WALLACE

Summary

Astronomers of all persuasions are now dependent on computers, and none more so than observational optical astronomers. It is clear that the rapidly increasing computer power available to such astronomers will find a rôle in on-line data handling (both in terms of data acquisition and instant data reduction) and come to be regarded as indispensible. In an attempt to explore some of the effects of increased computing power, I have taken the ASPECT application, which is still unique to the AAT, and examined how different it might be when implemented using modern computers and software.

1 Introduction

Optical astronomers were at one time very conservative about automation, mainly on reliability grounds. In the mid-1950s, for example, there were still lingering doubts about the wisdom of including electronic servo mechanisms in large telescopes (Felgett, 1956), and even in the late-1960s there was a serious suggestion that the planned Anglo-Australian Telescope (AAT) should be driven by falling weights rather than be at the mercy of the electricity supply. Perhaps reassured by the successful use of complex electronics and computers by radioastronomers, optical astronomers allowed themselves to be persuaded, and are now enthusiastic and innovative users of all these techniques.

Early uses of computers in optical work were for data gathering, instrument control, and telescope control, with one or two advanced pieces

287

of automation – for example the PDP-8 based Wampler-Robinson Image Dissector Scanner (Robinson & Wampler, 1972) – also offering some on-line data reduction capabilities. Astronomers now take for granted complete automation of the first three of these functions, and are becoming much more demanding about the fourth. It is the purpose of this paper to examine some of the effects of the recent rapid advances in computer hardware and software on optical data acquisition. To do this I am going to look at just one example – ASPECT, the AAT's area scanning spectrograph system. I have chosen this example because it requires comparatively large volumes of data to be handled, it is highly dependent on flexible detector and telescope control, and it uses substantial amounts of central processor (CPU) time. Full details of ASPECT can be found in Wallace (1985) and Clark *et al.* (1984).

2 ASPECT

ASPECT (which stands for Area SPECTrograph) is a software package which coordinates the acquisition of spectral data using the image photon counting system (IPCS) with telescope scanning to produce spectra of rectangular regions of the sky. The 3-dimensional data arrays it produces (called 'data cubes') have one wavelength dimension and (usually) two spatial dimensions – along and normal to the spectrograph slit.

Each data cube (see Fig. 22.1) can comprise up to 15 million 8-bit pixels (or, alternatively, 7.5 million 16-bit pixels), made up of 'windows' (2D spectra) up to 512K pixels in area, with the normal format limitations peculiar to the AAT IPCS and a few others arising from special timing considerations. A typical format would be 1600 spectral bins by 100 slit increments by 90 telescope positions. It is possible to record only specified bands of the spectrum (regions around lines of interest usually), and this allows economy of tape and data reduction time as well as greater spatial coverage. The basic 15 Mbyte limitation comes from the capacity of a 2400 ft 800 bpi tape.

2.1 Hardware

Figure 22.2 is a simplified diagram of the hardware. The detector (IPCS plus electronics) increments locations in a 1 Mbyte external memory

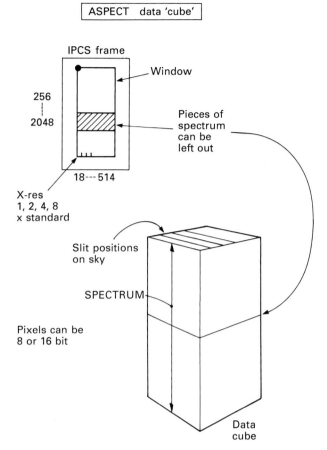

Fig. 22.1. The ASPECT data cube. Each ASPECT observation produces a 3D array of up to 15 million elements, consisting of a series of 2D spectra taken (usually) at different slit positions. A typical format is 1600 spectral bins by 100 slit increments by 90 telescope positions.

(X-mem), which is equipped with a hardwired CRT display. The instrument computer system (ICS) is a 64 Kbyte Interdata Model 70 (Perkin-Elmer 7/16). There is a 67 Mbyte Winchester disc for scratch storage, and 800 bpi 45 ips magnetic tape for offline storage. The IPCS itself (including the X-mem) is directly interfaced to the Interdata's I/O bus. The bus to the other instrumentation (the UTC register for instance) is Camac, which also provides the link to the Control Computer System (CCS–another 64 Kbyte Interdata 70). The data reduction and analysis computer is an 8 Mbyte VAX-11/780. Data reduction during observing is accomplished by physically

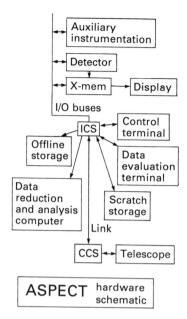

Fig. 22.2. A schematic of the ASPECT hardware showing the main components of the current ASPECT implementation. A new implementation of ASPECT would probably use a very similar scheme.

taking tapes from the Interdata to the VAX three floors below the control room.

2.2 Control dialogue

The ASPECT runs are set up and controlled by means a straightforward 'glass teletype' question-and-answer dialogue with the user. The questions are designed to default sensibly when the 'transmit' key alone is pressed. Together with a crude but effective facility for storing pre-entered sets of parameters, this approach allows each new run to be set up very rapidly and with a minimum of thinking.

The information supplied by the user consists of comments (pre-session, pre-run and post-run), whether the pixels are to be of 8 or 16 bits, the IPCS window parameters, the regions of spectrum to be saved (usually the whole spectrum is selected), the scan position angle and step size, whether the scan is to be disjoint (allowing a standard region of sky to be observed as well as the field of interest itself), the position of the scan (and the disjoint segment if used), the dwell and dead times, and the length and frequency of the periods of autoguiding.

2.3 Online displays

The online display facilities consist of the following:

The hardwired IPCS display shows the total integration so far for the current telescope position.

The run description, a run number, and the current scan and step number are displayed on the top line of the instrument control dialogue screen.

On another screen, a simple dialogue enables an uncalibrated spectrum to be shown on a graphics CRT, integrated across a given region of sky and range of wavelength bins. This allows spectral signal-to-noise ratio to be estimated; its most serious shortcoming is that 'end-on' views of the data cube (i.e. pictures of the sky) are not available.

2.4 Software

The ASPECT software consists essentially of two programs running in parallel under the RTOS operating system on the standard AAT Interdata Model 70 Instrument Computer (ICS), run after the IPCS has been set up using the normal IPCS software. The two programs are called ASCON and ASDATA respectively. ASCON is responsible for running the instrument; it contains an interrupt service routine (ISR) which is triggered by the IPCS frame flyback (FFB), and so consists of two independent but cooperating threads of processing. The main function of the ASDATA program is to write completed runs to tape, which can happen while the next observation is proceeding; it also provides a rudimentary data display facility. The whole system consumes 62 of the available 64 Kbytes of memory, as follows:

RTOS	28K
ASCON (including FFB ISR)	16K
ASDATA	17K
spare	2K
global common	1K

ASCON is heavily overlaid to conserve memory. Much of it is written in Fortran, except for the ISR and the overlay that controls the run itself, which are written in Assembler. ASDATA, also overlaid, is Fortran except for the innermost loop of the display function, which is coded in Assembler for speed. The total number of source lines in the whole ASPECT system is about 5000.

2.5 *Data management and telescope control*

An ASPECT session consists of a sequence of independent runs, each of which produces a 3-D data cube, output to tape as a FITS image after the run is complete. Each run consists of a sequence of IPCS exposures synchronised with telescope steps. The steps occur in a repeating cycle called a scan. The data cube consists of a series of 2-D pictures, one per telescope position. Each picture is the sum of all the exposures for that telescope position so far in the run. The data cube resides on the 67 Mbyte Winchester disc, and is up to 15 Mbyte in size. Two cubes can be stored at once to allow overlapped observing and writing to tape.

During an exposure, the detection of photon events by the IPCS causes locations within a particular region of the X-mem to be incremented. Successive exposures apply to different X-mem regions; the bias address which determines which region is to receive increments is changed by the ISR between exposures. The bias circulates round the whole of the X-mem and there is, in general, no fixed relationship between a given step and a particular region of the X-mem.

Outside the ISR, asynchronously, and fully overlapped with data taking, the data cube is circulated picture by picture through the X-mem to receive new exposures. Once an exposure has finished, the region of X-mem allocated to it is ready to be read from the X-mem and stored away on disc. This allows another picture, from somewhere else in the data cube, to be read from disc and written to the same region of the X-mem, where it will await its turn to be updated. As long as an exposure time has been chosen which is comfortably longer than the total X-mem to disc to X-mem swapout time, the X-mem is always kept fully stocked with pictures and no time is lost between one exposure and the next. The data swapout rate is a function of the IPCS window parameters, which if carefully chosen allow a throughput of up to 27 Kpixels sec^{-1} to be achieved.

Communication between ASPECT and the AAT control system takes place through the Camac Interprocessor Link, and consists of demands to perform telescope offsets and to control the autoguider. Each telescope step demand is despatched about 0.23 sec in advance of the time the step is needed, to compensate for the known delays in the system, so enabling almost continuous scanning with little dead time. Exposure times in practice are limited by data swapout rates; 2–5 sec is usual, with the telescope step one slit width, and 0.2 sec or less of dead time. Periodically, the telescope is brought back to a reference position and autoguiding is enabled for a few seconds. In between times the guider is 'frozen' and the telescope tracks open loop. An example of the open loop tracking is shown in Fig. 22.3, where the source is the Moon and involves the additional complication of non-sidereal tracking.

3 The impact of new hardware

Although ASPECT was developed comparatively recently (1983) most of the equipment used is 10–15 years old, and much more powerful hardware has since become available to astronomers. What differences might the astronomer expect to notice if an ASPECT implementation were done using the sort of equipment now being installed on large telescopes? Important factors to consider include software development costs (which affect how many such facilities can be provided and their sophistication), reliability, efficiency of operation, ease of use, the rate at which data can be acquired, and the quality of the on-line data evaluation facilities.

I will consider each hardware item in turn. I do not propose to consider detector developments, beyond assuming that increased data gathering capacity will remain a primary goal.

3.1 External memory

The current IPCS X-mem, *circa* 1976, has a capacity of 1 Mbyte. The speed of direct memory access (DMA) transfers to and from the Interdata 70 is about 0.5 Mbyte s^{-1}. There is a viewing CRT, with switch selection of contrast and region of window.

A new, general purpose, X-mem is planned for the AAT (Waller, 1985). It is based on VME bus and has its own 2 Mips 68020 processor, with floating point support, plus an image display. It would be possible to add an array processor. The transfer rate would be about 1 Mbyte s^{-1}. The initial system was to have a 6 Mbyte memory (not including the 68020 memory of course). In the year that has elapsed since the proposals were published, memory prices have dropped by a factor of ten, and so capacities like 64 Mbyte are feasible.

The main effect of these developments on a new ASPECT implementation would be that the whole data cube could reside in the X-mem for the entire run. This would greatly decrease the complexity of the data management software (no roll-out-roll-in) and eliminate I/O speed related format limitations. Telescope scanning could be as fast as the telescope allowed, and almost continuous sweeping, with improved immunity to variable transparency and seeing, would be usual. The largest format currently in use is about 7 million pixels; with a 64 Mbyte X-mem the size could go up to 16 million pixels, or 32 million if double buffering is used (to overlap writing to disc with the next run). Although the current roll-out-roll-in techniques would allow larger formats still, it is unlikely that the required processing power for offline data reduction and analysis will be available for some years; the current ASPECT formats bring even a dedicated VAX-11/780 to its knees. Another advantage of keeping the data cube X-mem resident would be that the powerful display and data manipulation facilities would allow specified parts of the data cube to be continuously monitored, a profound improvement over the current facilities. (Note that access by general purpose data analysis software of data objects in the course of an exposure will not necessarily be easy to arrange.)

3.2 Instrument and telescope computers

The *circa* 1973 Interdata 70s have 64 Kbyte memories (considered outrageously large by some noted experts of the time), of which 28 Kbyte is taken up by the RTOS operating system. The word length is 16 bits, and typical instruction times are 1–5 μsec (integer) and 50 μsec (single-precision floating point). Double-precision floating point is done in software, costing 1 msec or more per operation. The integer performance is quite respectable even by VAX and 68000 standards – simple register instructions take only 1 μsec, and multiply and divide are done in hardware. Any given algorithm will, of course, tend to require rather more instructions than on a modern machine, there being fewer addressing modes and only limited support of 32 bit integers. Floating point is not so competitive, and double precision is essentially prohibitive for online applications. Overwhelmingly the most severe limitation is memory, and much of the expertise developed in using these machines to the full was in memory conservation and overlaying techniques.

Though a dual processor system was chosen for the AAT, the difficulties experienced in implementing reliable and flexible link facilities, plus worries over reliability, precluded use of multiple processors for any given control application. These days off-the-shelf hardware and software have made it much easier to build multi-processor architectures, although the software is likely to be somewhat simpler if it can all be fitted into one processor. Memory sizes have increased enormously, with 8 Mbyte on a single board available at much less cost than an 8 Kbyte increment a few years ago. Operating systems have become much greedier, however, and it is not uncommon for the memory resident part to require 1 Mbyte or more. Modern micros have 32-bit architectures, or 16-bit architectures with extensive support of 32-bit data. Integer instruction times start at less than $0.5\,\mu$sec; single precision floating point instructions take from $3\,\mu$sec, and double precision is hardly any more expensive at $5\,\mu$sec. Some machines support floating point formats of even higher precision.

In a new implementation of ASPECT, the main improvements offered by modern computers would be the vastly increased memory, which would save much programming time and effort, and speed, which would allow much more use of high-level languages and floating point. However, any potential savings in software costs are likely to be nullified by the ever increasing demands made by the astronomers, who will come to expect much more attractive instrument control and better on-line data evaluation facilities.

On other control applications (especially telescope control) the advent of networking will make distributed computing feasible, greatly reducing the need for long distance wiring. An Ethernet local area network (LAN), requiring only a single coaxial cable, could take the place of hundreds of twisted pairs. Whether the overall complexity of the software would be reduced by spreading the computing load into distributed processors in this way is not yet clear, especially in applications requiring a high degree of synchronization between the devices being controlled.

3.3 Terminals

The terminals available on the AAT Interdata computers are alphanumeric VDUs, upper case only. They happen to be connected to the computers via very high speed DMA interfaces; this was done to enable instant display of menus, but they are most commonly used as 'glass teletypes' in order to avoid software complexity. On the ICS there is a storage CRT, interfaced via Camac and of 1024×1024 resolution. Again, the transfer rate is comparatively high.

Modern instrumentation and telescope computers have VDUs with lower case and clearer screens, and a variety of graphics displays, from raster line graphics devices to image displays. Single-user workstations may have an important rôle in this type of work, allowing multiple text and graphics windows to be arranged on a single high resolution screen. The workstations will also offer high speed compared with the serially interfaced terminals astronomers have become used to (which are, incidentally, much slower than the antique AAT screens).

In an ASPECT implementation, these facilities would enable a more pleasing and less tiring user interface than the current one. There could also be major gains in observing efficiency, especially if image displays can be used to mix sky data with graphics overlays. For example, the main TV picture could show the region to be scanned with a given set of parameters, plus a coordinate grid and instant readout of corrected RA/Dec for any nominated point. And during the run the current slit position could be shown on a stored picture of the field.

3.4 Interprocessor link

The ICS:CCS link on the AAT is an effective but idiosyncratic system implemented via the shared Camac. It offers certain unusual features: communication from the telescope computer is in the form of a 'noticeboard' rather than message-style, and multiple instrument computers can use the link at once. It is polled by the telescope computer, introducing a latency of 50 or 100 msec depending on what is being done. A throughput of a few Kbyte sec^{-1} could be achieved should any application require it.

Nowadays an Ethernet LAN would be used, in conjunction with standard software. A throughput of 100 Kbyte sec^{-1} would be available, and the delay between transmission and receipt of short messages would only be 100 μsec or so.

The impact of the new hardware on ASPECT itself would probably be small, assuming that all the required telescope control facilities had already been provided. Even with new hardware, applications which combine rapid and precise telescope scanning and data acquisition will not be easy to implement.

3.5 Scratch storage

ASPECT relies on the existence of a comparatively modern Winchester disc acquired by the AAO, *circa* 1980, and would have been severely limited in format and (to a lesser extent) speed had it been implemented on the original ICS/CCS discs. These are removable single platter devices, of 2.5 Mbyte capacity and about 0.1 Mbyte sec^{-1} transfer rate.

The Winchester disc used by ASPECT has a usable capacity of 67 Mbyte and a transfer rate of 1 Mbyte sec^{-1}.

A new instrument computer would have multiple Winchester discs, probably 500–1000 Mbyte in size, and still offering around 1 Mbyte sec^{-1} in practice.

If the X-mem was large enough to avoid continuously circulating the data cube through disc, the main effect of increased disc capacity might be to reduce the need to write tapes during observing. This could be a considerable advantage; tape mounting and labelling is a chore, there are opportunities for errors, and if a bad tape is encountered (not uncommon) telescope time is wasted while recovery procedures are carried out to get the data safely onto a new tape.

3.6 Offline storage

The current AAT ICS tape drives are 800 bpi 45 ips, offering sequential access to 20 Mbyte of data at about 0.03 Mbyte sec^{-1} and a media cost of 50 p per Mbyte for a full reel.

Current instrument computers still use magnetic tape, but at 6250 bpi. Tape speed is probably similar to the AAT drives, perhaps 50 ips, although more expensive drives are available at 2–3 times this speed. The capacity of a single reel is up to about 150 Mbyte. The transfer speed is 0.25 Mbyte sec^{-1} or so, and the media cost for a full reel is about 7p per Mbyte.

These developments are unlikely to have a particularly pronounced effect on ASPECT, because (i) tape writing is relatively infrequent and is overlapped with observing, and (ii) increases in tape density will be matched by increases in the rate of data acquisition.

In a few years it may be commonplace to return from an observing trip with one or two optical discs rather than a box of heavy tapes. The 12″ ones carry 1 Gbyte per side, are random access, and typically have transfer rates of 1 Mbyte sec^{-1}. The media cost is still rather high, at 23p per Mbyte. Even this figure assumes a completely full 2 Gbyte volume, suggesting that each astronomer will be issued with a single disc and only given another when he has filled the first up (like exercise books at school).

3.7 Instrumentation I/O bus

The AAT uses Camac, which offers somewhat more than 1 Mbyte sec^{-1} transfer rates (usually much less in practice).

Modern systems may instead opt for the VME architecture, for which many more devices are available, albeit comparatively complicated ones—processors, image displays, etc. The peak transfer rate is 40 Mbyte sec^{-1}, but typical devices rarely exceed a few Mbyte sec^{-1}.

ASPECT is not Camac-limited, and so improved facilities in this area would not have any effect, beyond the replacement of some highly specialised AAO Camac software with more industry-standard techniques. Other instrument applications will certainly benefit from the higher data bandwidth, given the trend towards ever larger formats, providing the detector itself has an acceptable readout time.

3.8 Data reduction and analysis

When ASPECT was implemented, the AAT VAX-11/780 was already available, and this played a vital rôle in data evaluation during observing. A major limitation is that data have to be transferred by physically carrying tapes from the AAT control room to the VAX room several floors below. It is reasonable to ask for perhaps 250 Mbyte of scratch space, and as long as the VMS page file is big enough, subsets of an observation can be examined by means of Starlink ASPIC (Disney & Wallace, 1982; Hartley & Lawden, 1986) routines, using a 512 × 512 image display.

Over the next few years, telescope users will come to expect good access to a VAX 8200/8300 class machine while they are observing, plus a dedicated micro-VAX equipped with an image display of perhaps 1280 × 1024 format

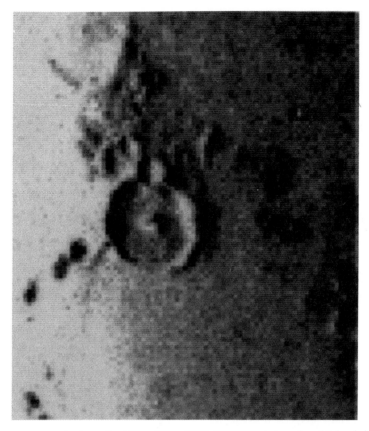

Fig. 22.3. A view of a $100 \times 100 \times 1600$ ASPECT data cube obtained during
testing. (One end of the spectrograph slit was vignetted.) The exposure was of
the Moon, and consisted of two complete scans totalling about 3 minutes of
integration. The telescope motions were the sum of the non-sidereal tracking
required to follow the Moon (applied directly via the telescope control system)
and the offset demands generated by ASPECT. This view of the data cube was
obtained by summing all the spectral bins for each spatial position, and is thus
simply a broadband picture of the Moon. For each pixel shown there is a solar
spectrum of 1600 bins.

and 500 Mbyte of disc space. Data will be transferred via a LAN rather than
tape, each ASPECT observation taking a minute or two to copy across.

The improvements in data reduction and analysis facilities will have a
large effect on ASPECT and most other forms of observing. The growth in
computer power will encourage increased quality and sophistication in
online analysis facilities, and for standard observing modes essentially fully
reduced and calibrated results will be available in real time. Note, however,
that VAX processors faster than the 11/780 are still rare and expensive–even

the 8300 mentioned above cannot run a single job faster than an 11/780, although its dual processor architecture gives a greater overall throughput– and so the processing times that UK astronomers have become used to since the inception of Starlink in 1980 are likely to remain the norm for some time. It is possible that fast attached processors will become available which can be efficiently programmed in Fortran without special techniques. If this happens, speed gains of at least 10 will be available, which could have a considerable impact on the popularity of 2D and 3D data acquisition.

3.9 Software

The AAT Interdata systems were advanced for their time in that punched cards could be used for program development rather than merely paper tape, and there was a disc on which subroutine libraries could be stored in object form. The RTOS operating system was very advanced, and included a relocating loader which allowed independent tasks to be run anywhere in memory. A Fortran compiler was available as well as an assembler. Debugging aids were more or less confined to memory dumps.

Since the late 1970s there have been profound improvements in program development and debugging tools, which have not only speeded up program coding by a large factor but have also encouraged better programming standards. The efficiency of the code produced by, in particular, the VAX compiler is very high, and recoding in assembler turns out to give little improvement in speed, even for 'bit-shuffling' routines. The arrival of VAX/VMS as a standard in astronomy has made transportation of software from site to site straightforward, and coupled with the availability of international networking has increased the possibilities for imported applications by orders of magnitude. There are now many more individuals than before capable of developing new astronomical software and support- ing applications written by others. In the future, much better data reduction and analysis software should be available, with more flexible and efficient applications controlled by powerful but easy to use command languages, and, for specialized observations, it will be normal for the astronomer to prepare his own data reduction and analysis facilities beforehand. This will be one benefit of the ADAM system (Chipperfield, 1986), which attempts to integrate instrument control, online data reduction, and offline analysis within a single software environment. The single most important objective of these software environments is to put the astronomer back in charge. Most of the application programs will be relatively primitive with well defined properties, and it will be up to the astronomer to 'wire' them together in the correct way to solve his specific problems; he will not be limited by pre-ordained menu options or sausage-machine processing, unless he wishes to be.

The increases in machine power and operating system complexity have not, however, been without cost. Access to the central processor at low level, crucial for ASPECT and many other instrument control applications, is much harder, and real time behaviour less predictable. Under the AAO's specially enhanced Interdata RTOS system, entry into a privileged and uninterruptible state is clean and fast, and the small variations in the time taken to enter this state are well understood. The interaction of operating system activities with the scheduling of tasks at different priorities is likewise understood. This leads to efficient yet reliable and flexible systems, with a minimum of assembler code and with unimpaired access to debugging tools and system utilities. This is not so easy to do with modern multiuser minicomputers such as the VAX, and for the best performance it will be necessary to adopt some standard supermicrocomputer configuration instead, based for example on 68000 series processors and VME bus. This is not desirable from the software point of view, where doing everything from instrument control to word processing on the same machine is most convenient, and it will also be necessary to master VAX/VMS as a real time system as well, even though its CPU architecture and context saving requirements will always make the VAX slower than a specialized real time computer.

An all too common arrangement is where the application programmer is cut off from the hardware by two layers of software written by other people. One layer, developed by the hardware engineer, ensures that the hardware can only do things the engineer has thought of, and probably masks the timing properties of the device. The other layer, developed by systems programmers, constrains access to specific software interfaces which have been designed for generality rather than speed and which also limit the scope for optimising the user interface. The best facilities for the implementor of a new ASPECT would be (i) dumb hardware, efficiently interfaced to a single computer, (ii) a software environment (including the vendor operating system) which offers standardised interfaces and access to ready made applications, and (iii) formal and correct procedures to bypass (ii) completely so that carefully written and fully debugged code can run in a privileged and interruptible way within (say) $200\,\mu$sec of an external event.

4 Summary

I have shown that in some future ASPECT implementation the new hardware and software would enable conspicuous improvements to be made. There are worries about firm real time control on computers with operating systems that like to be in charge, and no real hope that fewer programmers will be needed (the effort that used to be expended in memory and speed optimisation will go into fighting software environments and graphics packages, etc.). However, a much more agreeable user interface will be available, with lots to look at and some useful aids to setting up and monitoring the runs. Many more individuals will be able to write software useful for own online data reduction, and keen astronomers will be able to develop their own special facilities. Data formats will grow in size, but will be held back by the perennial problem of inadequate offline computing power. The reliability of new hardware is substantially better than the old equipment used with the current ASPECT – a major worry because of the large investment of telescope time before the tape is written. The availability of large amounts of disc space will also help, and tape writing can if desired be done the next day rather than during the night.

In short, a new ASPECT will permit somewhat larger formats and slightly faster scanning, will be significantly easier to use, and will suffer from fewer hardware and liveware failures. However, it would be unwise to expect it to be easier, on balance, to implement (because so much more will be expected of the online data evaluation facilities) and (being ultimately limited by the available signal) it may not, in normal conditions and with a well prepared observer, be much more productive astronomically than the current system.

References

Chipperfield, A.J. (1986). Starlink User Note 94, SUN/94.1.
Clark, D., Wallace, P., Fosbury, R. & Wood, R. (1984). *Q. Jl. Roy. Astr. Soc.*, **25**, 114.
Disney, M.J. & Wallace, P.T. (1982). *Q. Jl. Roy. Astr. Soc.*, **23**, 485.
Felgett, P.B. (1956). *Occ. Notes Roy. Astr. Soc.*, **3**, 18, 143.
Hartley, K.F. & Lawden, M.D. (1986). Starlink User Note 23, SUN/23.8.
Robinson, L.B. & Wampler, E.J. (1972). *Pub. Astr. Soc. Pacific*, **84**, 161.
Wallace, P. (1985). ASPECT Manual (3rd Ed.).
Waller, L. (1985). "A proposal for a large external memory system for the Anglo–Australian Telescope", Issue 2.0.

PART VI
CONCLUSION

23

Modern technology and its influence on astronomers

V. RADHAKRISHNAN

Friends:

I am delighted to be here with all of you to felicitate Hanbury on reaching 70 without accident in this increasingly dangerous world of ours. I am also very pleased to represent the country of his birth, where he has numerous friends and admirers. I bring him greetings from India on behalf of all of them.

Almost all the talks so far have been on innovations and advances in instruments and techniques given by people who were responsible for, or involved in, the development of these items. My own most original contribution to modern astronomical instrumentation was to convert a Swiss cuckoo clock to electrical operation and to build it into a 21 cm hydrogen line spectrometer. The bird would come out and wake me up at just those times when the sweeping local oscillator would pass a round number so I could write it on the chart paper. This was 30 years ago, and judging from more recent instruments that I have looked at very carefully the idea clearly did not catch on. Given this setback, I thought the most useful contribution I could make to this meeting is to raise a few philosophical questions, not covered by other speakers, related to advances in technology and instrumentation.

In the early days of any science, as I am sure so many here will attest, all of the equipment used for any investigation was completely transparent to the investigators. This is hardly surprising since they had almost certainly put the thing together themselves. You didn't have other experts who could do this for you, and of course you couldn't buy such equipment off the shelf. This developmental stage in any field is a good time to be in it, when as someone said 'second rate men can make first rate contributions'. One of numerous examples in this general category is the development of amateur or ham radio. There was a time when all but a few very rich or otherwise atypical hams built their own equipment and knew exactly how it worked. They were the ones who pushed techniques to higher and higher frequencies where no one else went or wanted to go, before the War and radar brought new and different players into the game. Today in most countries the average radio amateur just buys mass produced equipment, usually Japanese. These hams may find newer and cuter ways of talking to each other using microprocessors and television terminals, but I suspect that their technical contributions to radio science are not what is determining its development today.

As we have heard here, the four decades since the end of World War II witnessed a spectacular development in all branches of astronomy, some of which came into being about half way through this period. It is amazing how rapidly the cost of front line astronomical instruments went up. This forced the formation of a large number of national and international collaborative telescopes, observatories and establishments as the only way of equipping them with the most sophisticated instruments. The nice side of this is of course the bringing together of people from very different places in a common effort to study what Hanbury calls our one sky. But I have always felt that this relentless march towards bigger and more sophisticated machines is not all good, and that there is another side to it. I will try and explore this suspicion to see what if any evil there is in this instrumentation explosion. Many of my remarks are prompted by my personal reactions to the growth of radio astronomical instruments, but in some way they must apply to all other branches of astronomy, and to a greater or lesser extent to other branches of science too.

Particle physics is perhaps the first and also the best example of a total change with the growth of the field in the scale and character of the experimental work involved. Not only do you need an army now-a-days to carry out a single experiment, but the establishment required cannot, except in the case of the USA and the USSR, be afforded by any single country. I will get off this subject now in view of all the talk about Britain pulling out from CERN.

When the equipment becomes very complicated, the average user does not understand how it works. In fact, he cannot reasonably be expected to, since he is not an expert in all the technical fields involved. Because numerous others will also be using the equipment, it follows that he will not be allowed to mess with it. He is less likely therefore to learn about its hidden characteristics, much less find new ways to use it or improve upon it. The hardware is usually frozen at some very early stage. Even the software is too complicated to mess with except for experts. Fixed menu, not a la carte, and don't talk about going into the kitchen to make suggestions to the chef.

Such an increase in the complexity and sophistication of instruments may be considered inevitable and even natural when any particular field comes of age. The fact that a machine has to consist of many black boxes whose detailed working we don't understand is unavoidable but there may be a price to pay. This has to do with the nature of our environment on whose surface we just crawl around like ants. If the environment is *nature itself* with all its beauty and mystery it evokes a desire to understand it, and is the best kind of scientific motivation. But if it is man-made and consists of black boxes fixed by servicemen when they go wrong, it has in general the opposite effect. I think such an environment blunts the senses, dulls the mind, deadens curiosity and tends to make us more moronic than we started out.

Good examples of such modern complicated instruments are large synthesis radio telescopes, which reduce the user to the status of the clicker of a camera. I believe you can even do this by telephone, and a completely processed map comes out of the system like a photograph out of a polaroid camera. I had one some years ago, and was being frustrated in the attempt to understand its working by my four year old son. He kept wanting to take pictures with it everytime he saw me with the machine. Late at night when he was asleep, I figured out its mechanical working, and loaded the magazine somewhat differently. The next day when he saw me with the camera he asked for it, and I suggested he should take a picture of me. He pointed at me, and clicked, and the camera made the usual noises, and out came a picture, but it was of his *mother*. He took one long hard look at the picture, handed me back the camera, and never asked for it again. I could never hold down a job at Westerbork or the Very Large Array (VLA) because it would only be a question of time before I figured out how to give users each others' maps, suitably labelled of course.

On the subject of maps, let me vent my spleen a little. A proper radio map represents in a sense a total description of the region of sky at that frequency, and therefore all that can be gleaned from here. They were therefore a dream before aperture synthesis techniques made high resolution maps available, and one believed, or at least hoped, that given such pictures of radio sources, their inner workings would be laid bare. I remember the excitement I felt when Ryle showed the first maps made at Cambridge, and there is no question that much enlightenment has resulted from such pictures although not in proportion to their availability now. Once it was clear how maps could be made, they became compulsory and everybody had to try making them. Then they became computer art, and now instead of being revelations of the secrets hidden in distant parts of the universe, they are becoming a bit of a bore at meetings. We all remember when the marvellous technological invention of colour photography became available to the man in the street, how travel slides used to be inflicted on one often with little concern for what was in the picture. The same thing is happening now with synthesis telescopes, even more marvellous instruments, from which 'coloured' pictures are easily available for the asking.

Talking about large synthesis telescopes, I cannot help relating a VLA story that I have first hand from a typical user, in this case a solar radio astronomer. He was staring into his terminal for hours and complaining at not seeing what he expected to. Receiver and other experts came in, did various checks and also failed to get any enlightenment from the terminal. Finally it was suggested that the system must have broken down calling for a major investigation. Barry Clark who happened to walk in at that moment goes over to the window, peers through a pair of binoculars at the shadow of a feed on the reflecting surface, tells the user that it would help if the telescopes were pointed at the Sun and walks out. I don't need to spell out the moral to be drawn from this story.

It is well known that students who have participated in the building and commissioning of an instrument before using it have a well rounded scientific training which gives them maturity and a confidence that is often missing in later generation students. Readymade sophisticated instrumentation may produce high quality data but 'homemade' equipment might be more conducive to good scientific thinking. One moral from all of this could be that the good guys should move on to a new and unexplored field which has a long way to go before becoming a bandwagon and subsequently a rat race. I don't know how many scientists think in these terms, but I am certain that new fields are not discovered and developed by pushers of buttons on black boxes.

One is really talking about the phenomenon that Charlie Chaplin addressed so poetically in his film *Modern Times*. Like everyone else I thought it was hilariously funny when I saw it as a child. I still laugh every time I see it, but never when I think about it. There must be many who thought at the time the film was made that the dehumanisation associated with the transition from craftsman to assembly line worker in a factory was grossly exaggerated just to make it funny. Most thinking people now are horrified by what this transition has done to the modern world, but they think

Fig. 23.1. Modern Times: Modern Astronomy.

of scientists as very different, and very creative creatures with free minds. In my view, *Modern Times* came to science quite a long time ago, but never so clearly as with the advent of the big computers (Fig. 23.1).

Modern computers are among the cleverest inventions of the human mind, but the computerization of the world must surely be the work of the devil. Every time I walk into the terminal room of a large computing centre, I cannot help feeling that I am in a slot machine arcade. I see row upon row of the same mindless stare occasionally relieved by a smile in the one case when some money drops out and in the other case when something works. I have a modest illustration of such a terminal case which shows a typical modern astronomer at work (Fig. 23.2). It is applicable to astronomy at all wavelengths, not to mention all other fields of research which have reached this level of sophistication.

If I may quote Sir Bernard Lovell on the use of computers, 'the very fact that you have a major institution completely computerized means that everybody has to use these systems. It is extremely unlikely that proposals for work will be accepted which do not use the major facilities of an

Fig. 23.2. The typical modern astronomer, a terminal case.

establishment. Therefore, this is directing the work of every observatory into narrower and narrower 'channels'. When the channels are narrowed you will find, or not find, only what you are looking for. For it to be significant if you do, it helps to be a genius. Most discoveries of importance in astronomy were made not by restricting the channels but by scanning the output so to speak with the human brain.

A whole conference was devoted to discussing just this matter of unplanned discoveries a few years ago at Green Bank. I was not present, but from the Proceedings I see that there were many there who expressed just the concerns I am voicing here, including the bit I just quoted.

The data output rate of modern machines is so high that even a short run will take a long time to process, let alone assimilate its meaning. If it were only the speed of the machine we were talking about, it would be as silly to complain about this as about a fast camera with needs only 100th of a second instead of 10 seconds while everybody is holding their breath. The speed allows observations to be programmed like tube trains in the rush hour so you get the most passenger miles per unit of time. Whether the output of significant results keeps step with the data rate or not, the large investment dictates that you keep the machine running flat out. But as the demand for time is even greater than that available, proposals must necessarily be reviewed.

Hanbury is on record as saying that a logical committee reviewing innovations would have rejected the piano as unworkable and that anybody given drawings of a bicycle would reject it at once. He was making the point that most innovations on which our civilization depends now would not have been supported by a committee of review. I agree with him. In the present context we should ask about the astronomical discoveries on which our present notions of the galaxy and the universe really rest. Would proposals for these observations have passed review committees, and how many did. Or were most of these discoveries made at places where the committee approach and peer review were pushed aside by individuals with both power and judgement.

Another but less widely appreciated evil associated with modern instrumentation is the attitude of those using such equipment to earlier and often pioneering work done elsewhere with more primitive apparatus. I know of more than one case where an important discovery becomes attributed to those who did not have the wit to look for it in the first place, but who were able to produce good looking data by raking already broken ground with a bigger and better instrument. Our Soviet colleagues are particularly prone to be victims of this phenomenon at the hands of the West.

Going beyond data processing, there is another use of big computers in astronomy which is fraught with pitfalls. Complicated astronomical phenomena are often modelled using a Monte Carlo or other favourite casino method. When an output is obtained which roughly matches the observations, this is taken most seriously as proof that the natural process involved was as in-putted to the computer. The notion that uniqueness is essential before any such modelling can be taken seriously does not seem to be widely appreciated.

I will try to be a little constructive in the last few minutes, but let me give one more example on the negative side. In the selection of problems for study, elegance and good taste are frequent casualties when big computers are around. Instead of using them to solve problems whose importance was always recognized, but where progress could not be made because of the necessary computation involved, there are more and more of those who choose problems which require enormous calculating power as if that were a virtue in itself. Such types identify themselves by telling you proudly how much Cyber or Cray time was used to obtain their inconclusive or otherwise meaningless results.

Leaving out all the nasty remarks, the useful part of what I have been trying to say can be summarized as follows.

LARGE
ASTRONOMICAL INSTRUMENTS/COMPUTING FACILITIES
Obvious gain in sensitivity, speed, coverage, labour saving,
range of projects made possible – but leads to:

The black box syndrome characterized by complexity. Here one has to distinguish between (1) the complexity of how something is done (which is inevitable) and (2) the complexity of what is done. A good example of the first kind would be the processing and analysis of the data obtained from the binary pulsar PSR1913 + 16. I cite that because I believe that even if the discovery of pulsars might have been delayed for years by the use of computers in the earliest observations, the only evidence we have for gravitational wave radiation simply could not have been obtained without the use of computers. Whatever be the details of the programme used, we know what they finally did, like packages for things like evaluating integrals. A hardware analogue would be a correlator which inside has all kinds of tricks, bells, whistles, idiosyncracies and personal touches, but gives a correlation (as normally understood) as output. It is unexceptionable and indeed commendable.

An example of the other kind of complexity is 'image processing' which is not such a routine operation. Here we have hidden inputs, assumptions, biases, etc., which the user does know about or tends to forget. A central core – the 'priesthood' – knows the origins and motivations for all of this and others have to – or choose to – accept their results uncritically. Standardized 'packages' decide what is good for you, and you can sleep easily if you share that belief. If you don't, you could try writing your own and will soon discover the man decades that have gone into them and the madness of devoting a good fraction of your life to a similar exercise. If you persist, you will lose sight of what you wanted to do in the first place, but get a well paid and highly respected job directing the production of bigger and better packages. The net result, and an insidious one, is that the existence or availability of packages determines what is done and what is not.

Not surprisingly scientific papers describing work in which such packages are used read like endorsements for various proprietary and patent medicines e.g. – 'were done using AIPS as implemented on Startrek under VMS…' instead of – 'the image was processed using Hogbom's CLEAN method

3213. *May 24, 1986*   Minor Janitorial Duty   *Kerry*

Changed: IMIO, IMPFIT, TVCUB, TVSLD and TVHXF. See #3080.
Moved to 15JUL86 this date.

3214. *May 27, 1986*   I*4 in calls to YIMGIO   *Kerry*

Several AP tasks contained an I*4 variable as the NPIX argument in calls to YIMGIO including APCLN, MX, VM and VTESS. These caused TV displays to fail on the Convex. A bug was fixed in VTESS. See #3081 for details.
Moved to 15JUL86 this date.

3215. *May 27, 1986*   Misc   *Eric*

Removed TAB characters from PCNTREQ.INC, YDEA.INC, COMPILE.COM, CREADIR.COM, CREATOLB.COM, and OP-TIONS.COM, Deanza routines YCRCTL, YGGRAM, YGRAPH, YMKCUR, YSPLIT, and YTVCIN, and M75 versions YGRAFE and YGRAPH. Fixed long lines in CONDAT.INC, PCNTREQ.INC, and YDEA.INC. Removed blank lines from YCRCTL (YDEA), and Comtal (V20) versions of YCMSET, YLUT, YOFM, and YSTCUR.
Moved to 15JUL86 this date, nowhere else.

Fig. 23.3. An extract from the AIPS (astronomical image processing system of the NRAO) newsletter.

Another method for finding suspicious data is provided by the task FFT. Transform your image back into the *(u,v)* plane by running FFT and then display the results on the TV. Verbs like CURVALUE and IMPOS will help you find the *u* and *v* values for abnormally high cells. Then UVFND with OPCODE 'UVBX' will print the data surrounding these cells and UVFLG can be used to delete the bad data. This method is particularly effective on the residual images from CLEAN. (You can instruct APCLN and MX to put out a residual image by setting BMAJ less than 0.)

5.4. Banana poundcake

1. Mix in large bowl until blended:

   $1\frac{1}{3}$ cups mashed bananas (4 medium)
   1   pkg. ($18\frac{1}{2}$ oz.) yellow cake mix
   1   pkg. ($3\frac{3}{4}$ oz.) instant vanilla pudding mix
   $\frac{1}{3}$ cup salad oil
   $\frac{1}{2}$ cup water
   $\frac{1}{2}$ teaspoon cinnamon
   $\frac{1}{2}$ teaspoon nutmeg
   4   eggs at room temperature

2. Beat at medium speed for 4 minutes.
3. Turn batter into greased and lightly floured 10-inch tube pan.
4. Bake in 350° oven for 1 hour or until cake tester inserted in cake comes out clean.
5. Cool in pan 10 minutes, then turn out onto rack and cool completely.
6. If desired, dust with confectioners sugar before serving.

   Thanks to the United Fresh Fruit and Vegetable Association.

Fig. 23.4. Another AIPS newsletter extract.

which emphasises point sources, but incorporating Steer's modification to avoid oversubtraction in extended regions of emission...'.

I am a grateful recipient of periodic update information on these packages some of which we use. I have a couple of slides showing portions of two out of 200 or so pages of the last newsletter picked at random. The first example (Fig. 23.3) is typical of the whole lot and illustrates how 'easily' a user can keep up with what is really being done to the programme! But to my surprise, the second can be implemented even without a VAX computer (Fig. 23.4).

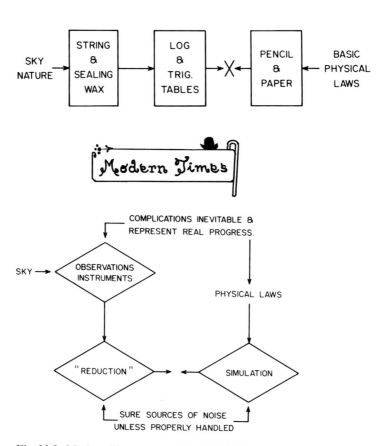

Fig. 23.5. Modern Times versus Good Old Days.

Let me now turn to the other kind of package, usually bigger, like the 'hydro code' or the '*n*-body code' which is used to come up with models based on some input physics. The tendency here is:

(i) to forget the (usually simplified) nature of the input and ascribe wider validity to the results.

(ii) to attribute a uniqueness to the 'fit' which is usually not justified.

(iii) to ignore the limitations of the code itself (of which the 'priesthood' is aware).

Let me give you a very good example which relates to one of the most important problems in astrophysics, namely the supernova phenomenon. It has been found that in most core bounce calculations the outgoing shock stalls – there is no supernova explosion! Recently Wilson obtained a 'delayed explosion' of fairly low energy: $E \sim 10^{50}$ ergs. This is the first explosion to be found in numerical calculations. But Arnett, who is one of the *highest* priests in this business, repeated this calculation and did *not* find any explosion! The basic reason why different workers *disagree*, as Arnett himself pointed out and emphasized is that the errors in the calculations $\Delta E \sim 10^{50}$!!

> (iv) Even if (i) to (iii) above are overcome, one must be careful not to confuse success with understanding of what is going on. After all, a faithful simulation of a very complex system is likely to behave in as complex a fashion as the system itself.

Science is based on different experimenters agreeing with each other. But in both very large scale data processing and large scale simulation, the likelihood of independent efforts confirming each other in different ways is receding. An example of this is the binding energy calculation for the ions on the surface of pulsars. After going back and forth for years we are apparently back to square one and in total ignorance of what the surface of Neutron stars is really like.

So what does it all mean?

The relatively simple situation of the good old days has given way irrevocably to the present situation as illustrated in Fig. 23.5. But what of the future? What can we hope for in the GOOD NEW DAYS?

Instruments: Will undoubtedly get bigger and more complex – the people making them are clearly having a good time because it strains their capabilities to their limits. *They* are being creative. The only thing we should be asking is:

> (i) Are we building them simply because it is possible to do so like people scaling mountains because they are there. And are we building more and more of them for the same reason that Kurt Mendelssohn gives for the proliferation of the pyramids. There is less and less standing room at the top as the pyramid nears completion, and the best place for the slaves is at the base of a new pyramid. Projects like Space Sub-millimetre VLBI certainly appear that way to me.
>
> (ii) What are we going to do with the output?

Data analysis and reduction: No question that a large amount of software will be involved but one could hope for:

(i) More structured software with boxes performing standard well defined functions – boxes whose internal structure one can forget, leaving time and effort and discretion for what the user does with the boxes. Put differently, stop treating the user as if he were an idiot in scientific matters (e.g. aperture synthesis) and a wizard in computing matters.

(ii) More innovative software to look at the final result in different and interesting (not just pretty) ways, as was done by Ekers and Allen for Westerbork data. But perhaps going much further into pattern recognition, and what is known as 'artificial intelligence.'

Neither of these two characterisations of 'good new' software is really new. There are voices in the computer wilderness who have been trying to do all these things for years. Since we need them so badly in astronomy, it is quite fair to ask that we contribute as much to these fields (pattern recognition, artificial intelligence, higher level languages...) as in the past to receivers, detectors, etc. This is a creative, non-routine thing. Productivity in the multiplicative sense may go down but in the additive sense will go up. (We don't have to be like the three boys scouts who all helped the old woman across the street. For those who don't remember the story, it took all the three boy scouts because the old woman didn't want to cross the street in the first place.)

Much of this applies to the large scale simulations. In the new era we may hope that it isn't the development of/monopoly over a 'code', but the creative use of these to generate insight, not just producing numbers, as Hamming tried to tell us. Editors, review committees, etc., all have to take the responsibility for elevating the level at which our minds are operating. Otherwise, all of this effort will result in destroying the most marvellous of all instruments, the human brain.

I started with Hanbury, so let me end with him.

I hope I have persuaded you that there is a nasty side to the way we are going which makes the man we are honouring here the perfect example of an endangered species facing extinction. The simplicity and clarity of thinking and elegance of approach which have characterized all of Hanbury Brown's work and which we admire, are just those qualities in the pursuit of science which are threatened today by modernization, as is our very survival by the bomb, an excellent example of modern technology. Of course, it doesn't have

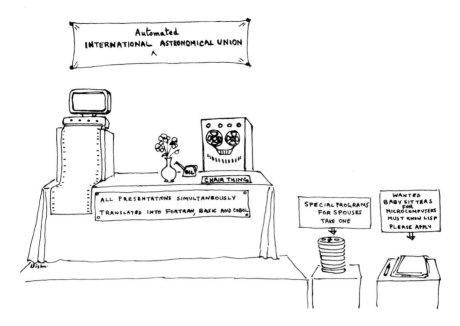

Fig. 23.6. Modern Times: new style I(A)AU meetings.

to be so. But it will if we do not start thinking about the meaning of what we are doing.

Those of you who were at the Delhi General Assembly of the IAU last year would have enjoyed the wit and wisdom Hanbury brought to even the most formal of business meetings, making them a pleasure to attend. Modernization may put an end to that too, and my last picture anticipates what future such meetings may be like, perhaps, already at Baltimore in 1988 (Fig. 23.6).

I thank you for your patience.

I am very grateful to. R. Nityananda for almost all of the constructive part of my criticism and to C.V. Vishveshwara for protraying so skillfully the situation as I see it now.

24

Concluding remarks

R. HANBURY BROWN

Some years ago I made the long journey from Australia to England to attend a one-day meeting on the instruments proposed for the Anglo–Australian Telescope. I remember that meeting very well because, having come about 12 000 miles, I lost my way between the hotel in Cambridge and the Institute of Astronomy so that I arrived late, after the opening speaker, a well-known theoretician, had started to talk. I was just in time to see him draw a flow diagram on the blackboard in which the chief engine of progress in astronomy was shown as being theory. Theory, so he said, leads to observation which produces data which in turn are compared with theory, and so knowledge progresses. At the end of the talk someone, it might have been Peter Fellgett, got up one second before I did and pointed out that a more realistic view is that the development of new instrumentation has a greater claim than theory to be the chief engine of progress.

Indeed, most of the great steps forward in astronomy are marked by the introduction of new instruments: the telescope, the camera, the photometer, the spectroscope, and more recently by the electronic image tube and by ultra-violet and X-ray telescopes which make possible the extension of the spectrum by observations in space; we may very soon see further steps marked by the introduction of the Hubble telescope and perhaps by very large ground-based telescopes which exploit servo-mechanisms and active optics.

Because the progress of astronomy, indeed of all science, depends fundamentally on the development of new instruments – on new ways of looking at the world – it is important to maintain close links between the observers and the development of new techniques. Optical astronomy used to be peculiarly weak in that respect; observatories were located up mountains, far from the laboratories where the new techniques, such as

317

electronics, were being developed and used. Many optical astronomers were, of course, expert in optics and photography but very few had any knowledge or interest in electronics, a subject which was soon to transform their science. Nowadays, astronomers can hardly imagine life without electronics, without their image tubes and computers – indeed many of them can talk of little else.

Radio astronomy has played, I think, a major part in demonstrating the importance of a close connection between instrumental development and observational astronomy. Radio astronomers did not need to isolate themselves on mountain tops far from the physicists and electrical engineers on whom their progress depended. In fact, it was this close connection with electronic laboratories that made radio astronomy blossom so vigorously and so fast. When I started work at Jodrell Bank in 1949, many of my colleagues knew very little about radio astronomy but they knew quite a lot about electronics; I suppose that nowadays, things would be the other way round. You cannot expect a modern astronomer to have an expert knowledge of all the techniques on which his work depends – that is one of the self-limiting factors in the progress of science and is part of the price we pay for that progress. This is why it is even more important nowadays to maintain and strengthen the links between observational astronomy and instrumental development. A conference like this serves to underline that point and in doing so, performs a valuable service to astronomy.

In conclusion, I would like to express my thanks, and I am sure your thanks as well, to the Director of the R.G.O., Alec Boksenberg, to Jasper Wall, to Beryl Andrews and to all the many other members of the staff who have made this excellent meeting possible.

Index

319